Machine Learning on Kubernetes

A practical handbook for building and using a
complete open source machine learning platform
on Kubernetes

Faisal Masood

Ross Brigoli

BIRMINGHAM—MUMBAI

Machine Learning on Kubernetes

Publishing Product Manager: Dhruv Jagdish Kataria

Senior Editor: David Sugarman

Content Development Editor: Priyanka Soam

Technical Editor: Devanshi Ayare

Copy Editor: Safis Editing

Project Coordinator: Farheen Fathima

Proofreader: Safis Editing

Indexer: Manju Arasan

Production Designer: Nilesh Mohite

Marketing Coordinators: Shifa Ansari, Abeer Riyaz Dawe

First published: June 2022

Production reference: 1190522

Published by Packt Publishing Ltd.

Livery Place

35 Livery Street

Birmingham

B3 2PB, UK.

ISBN 978-1-80324-180-7

www.packt.com

"To my daughter, Yleana Zorelle – hopefully, this book will help you understand what Papa does for a living."

Ross Brigoli

"To my wife, Bushra Arif – without your support, none of this would have become a reality"

Faisal Masood

Contributors

About the authors

Faisal Masood is a principal architect at Red Hat. He has been helping teams to design and build data science and application platforms using OpenShift, Red Hat's enterprise Kubernetes offering. Faisal has over 20 years of experience in building software and has been building microservices since the pre-Kubernetes era.

Ross Brigoli is an associate principal architect at Red Hat. He has been designing and building software in various industries for over 18 years. He has designed and built data platforms and workflow automation platforms. Before Red Hat, Ross led a data engineering team as an architect in the financial services industry. He currently designs and builds microservices architectures and machine learning solutions on OpenShift.

About the reviewers

Audrey Reznik is a senior principal software engineer in the Red Hat Cloud Services – OpenShift Data Science team focusing on managed services, AI/ML workloads, and next-generation platforms. She has been working in the IT Industry for over 20 years in full stack development relating to data science roles. As a former technical advisor and data scientist, Audrey has been instrumental in educating data scientists and developers about what the OpenShift platform is and how to use OpenShift containers (images) to organize, develop, train, and deploy intelligent applications using MLOps. She is passionate about data science and, in particular, the current opportunities with machine learning and open source technologies.

Cory Latschkowski has made a number of major stops in various IT fields over the past two decades, including high-performance computing (HPC), cybersecurity, data science, and container platform design. Much of his experience was acquired within large organizations, including one Fortune 100 company. His last name is pronounced Latch - cow - ski. His passions are pretty moderate, but he will admit to a love of automation, Kubernetes, RTFM, and bacon. To learn more about his personal bank security questions, ping him on GitHub.

Shahebaz Sayed is a highly skilled certified cloud computing engineer with exceptional development ability and extensive knowledge of scripting and data serialization languages. Shahebaz has expertise in all three major clouds – AWS, Azure, and GCP. He also has extensive experience with technologies such as Kubernetes, Terraform, Docker, and others from the DevOps domain. Shahebaz is also certified with global certifications, including AWS Certified DevOps Engineer Professional, AWS Solution Architect Associate, Azure DevOps Expert, Azure Developer Associate, and Kubernetes CKA. He has also worked with Packt as a technical reviewer on multiple projects, including *AWS Automation Cookbook*, *Kubernetes on AWS*, and *Kubernetes for Serverless Applications*.

Table of Contents

3
Exploring Kubernetes

Part 2: The Building Blocks of an MLOps Platform and How to Build One on Kubernetes

4
The Anatomy of a Machine Learning Platform

5
Data Engineering

6
Machine Learning Engineering

7
Model Deployment and Automation

Part 3: How to Use the MLOps Platform and Build a Full End-to-End Project Using the New Platform

8

Building a Complete ML Project Using the Platform

9

Building Your Data Pipeline

10
Building, Deploying, and Monitoring Your Model

11
Machine Learning on Kubernetes

Index

Other Books You May Enjoy

Preface

Machine Learning (**ML**) is the new black. Organizations are investing in adopting and uplifting their ML capabilities to build new products and improve customer experience. The focus of this book is on assisting organizations and teams to get business value out of ML initiatives. By implementing MLOps with Kubernetes, data scientists, IT operations professionals, and data engineers will be able to collaborate and build ML solutions that create tangible outcomes for their business. This book enables teams to take a practical approach to work together to bring the software engineering discipline to the ML project life cycle.

You'll begin by understanding why MLOps is important and discover the different components of an ML project. Later in the book, you'll design and build a practical end-to-end MLOps project that'll use the most popular OSS components. As you progress, you'll get to grips with the basics of MLOps and the value it can bring to your ML projects, as well as gaining experience in building, configuring, and using an open source, containerized ML platform on Kubernetes. Finally, you'll learn how to prepare data, build and deploy models quickly, and automate tasks for an efficient ML pipeline using a common platform. The exercises in this book will help you get hands-on with using Kubernetes and integrating it with OSS, such as JupyterHub, MLflow, and Airflow.

By the end of this book, you'll have learned how to effectively build, train, and deploy an ML model using the ML platform you built.

Who this book is for

This book is for data scientists, data engineers, IT platform owners, AI product owners, and data architects who want to use open source components to compose an ML platform. Although this book starts with the basics, a good understanding of Python and Kubernetes, along with knowledge of the basic concepts of data science and data engineering, will help you grasp the topics covered in this book much better.

What this book covers

Chapter 1, Challenges in Machine Learning, discusses the challenges organizations face in adopting ML and why a good number of ML initiatives may not deliver the expected outcomes. The chapter further discusses the top few reasons why organizations face these challenges.

Chapter 2, Understanding MLOps, continues building on the identified set of problems from *Chapter 1, Challenges in Machine Learning,* and discusses how we can tackle the challenges in adopting ML. The chapter will provide the definition of MLOps and how it helps organizations to get value out of their ML initiatives. The chapter also provides a blueprint on how companies can adopt MLOps in their ML projects.

Chapter 3, Exploring Kubernetes, first describes why we have chosen Kubernetes as the basis for MLOps in this book. The chapter further defines the core concept of Kubernetes and assists you in creating an environment where the code can be tested. The world is changing fast and part of this high-velocity disruption is the availability of the cloud and cloud-based solutions. This chapter provides an overview of how the Kubernetes-based platform can give you the flexibility to run your solution anywhere.

Chapter 4, The Anatomy of a Machine Learning Platform, takes a 1,000-foot view of what an ML platform looks like. You already know what problems MLOps solves. This chapter defines the components of an MLOps platform in a technology-agnostic way. You will build a solid foundation on the core components of an MLOps platform.

Chapter 5, Data Engineering, covers an important part of any ML project that is often missed. A good number of ML tutorials/books start with a clean dataset, maybe a CSV file to build your model against. The real world is different. Data comes in many shapes and sizes and it is important that you have a well-defined strategy to harvest, process, and prepare data at scale. This chapter will define the role data engineering plays in a successful ML project. It will discuss OSS tools that can provide the basis for data engineering. The chapter will then talk about how you can install these toolsets on the Kubernetes platform.

Chapter 6, Machine Learning Engineering, will move the discussion to the model building tuning and deployment activities of an ML development life cycle. The chapter will discuss providing a self-service solution to data scientists so they can work more efficiently and collaborate with data engineering teams and fellow data scientists using the same platform. It will also discuss OSS tools that can provide the basis for model development. The chapter will then talk about how you can install these toolsets on the Kubernetes platform.

Chapter 7, Model Deployment and Automation, covers the deployment phase of the ML project life cycle. The model you build knows the data you provided to it. In the real world, however, the data changes. This chapter discusses the tools and techniques to monitor your model performance. This performance data could be used to decide whether the model needs retraining on a new dataset or whether it's time to build a new model for the given problem.

Chapter 8, Building a Complete ML Project Using the Platform, will define a typical ML project and how each component of the platform is utilized in every step of the project life cycle. The chapter will define the outcomes and requirements of the project and focus on how the MLOps platform facilitates the project life cycle.

Chapter 9, Building Your Data Pipeline, will show how a Spark cluster can be used to ingest and process data. The chapter will show how the platform enables the data engineer to read the raw data from any storage, process it, and write it back to another storage. The main focus is to demonstrate how a Spark cluster can be created on-demand and how workloads could be isolated in a shared environment.

Chapter 10, Building, Deploying, and Monitoring Your Model, will show how the JuyterHub server can be used to build, train, and tune models on the platform. The chapter will show how the platform enables the data scientist to perform the modeling activities in a self-serving fashion. This chapter will also introduce MLflow as the model experiment tracking and model registry component. Now you have a working model, how do you want to share this model for the other teams to consume? This chapter will show how the Seldon Core component allows non-programmers to expose their models as REST APIs. You will see how the deployed APIs automatically scale out using the Kubernetes capabilities.

Chapter 11, Machine Learning on Kubernetes, will take you through some of the key ideas to bring forth with you to further your knowledge on the subject. This chapter will cover identifying use cases for the ML platform, operationalizing ML, and running on Kubernetes.

To get the most out of this book

You will need a basic working knowledge of Kubernetes and Python to get the most out of this book's technical exercises. The platform uses multiple software components to cover the full ML development life cycle. You will need the recommended hardware to run all the components with ease.

Software/hardware covered in the book	Operating system requirements
Kubernetes, Python, Spark, MLflow, Seldon, Airflow	Windows, macOS, or Linux

Running the platform requires a good amount of compute resources. If you do not have the required number of CPU cores and memory on your desktop or laptop computer, we recommend running a virtual machine on Google Cloud or any other cloud platform.

If you are using the digital version of this book, we advise you to type the code yourself or access the code from the book's GitHub repository (a link is available in the next section). Doing so will help you avoid any potential errors related to the copying and pasting of code.

A good follow-up after you finish with this book is to create a proof of concept within your team or organization using the platform. Assess the benefits and learn how you can further optimize your organization's data science and ML project life cycle.

Download the example code files

You can download the example code files for this book from GitHub at `https://github.com/PacktPublishing/Machine-Learning-on-Kubernetes`. If there's an update to the code, it will be updated in the GitHub repository.

We also have other code bundles from our rich catalog of books and videos available at `https://github.com/PacktPublishing/`. Check them out!

Download the color images

We also provide a PDF file that has color images of the screenshots/diagrams used in this book. You can download it here: `https://static.packt-cdn.com/downloads/9781803241807_ColorImages.pdf`.

Conventions used

There are a number of text conventions used throughout this book.

`Code in text`: Indicates code words in text, database table names, folder names, filenames, file extensions, pathnames, dummy URLs, user input, and Twitter handles. Here is an example: "Notice that you will need to adjust the following command and change the `quay.io/ml-on-k8s/` part before executing the command."

A block of code is set as follows:

```
docker tag scikit-notebook:v1.1.0 quay.io/ml-on-k8s/scikit-notebook:v1.1.0
```

When we wish to draw your attention to a particular part of a code block, the relevant lines or items are set in bold:

```
gcloud compute project-info add-metadata --metadata enable-
oslogin=FALSE
```

Bold: Indicates a new term, an important word, or words that you see onscreen. For instance, words in menus or dialog boxes appear in **bold**. Here is an example: "The installer will present the following **License Agreement** screen. Click **I Agree**."

> **Tips or Important Notes**
> Appear like this.

Get in touch

Feedback from our readers is always welcome.

General feedback: If you have questions about any aspect of this book, mention the book title in the subject of your message and email us at customercare@packtpub.com.

Errata: Although we have taken every care to ensure the accuracy of our content, mistakes do happen. If you have found a mistake in this book, we would be grateful if you would report this to us. Please visit www.packtpub.com/support/errata, selecting your book, clicking on the Errata Submission Form link, and entering the details.

Piracy: If you come across any illegal copies of our works in any form on the Internet, we would be grateful if you would provide us with the location address or website name. Please contact us at copyright@packt.com with a link to the material.

If you are interested in becoming an author: If there is a topic that you have expertise in and you are interested in either writing or contributing to a book, please visit authors.packtpub.com.

Reviews

Please leave a review. Once you have read and used this book, why not leave a review on the site that you purchased it from? Potential readers can then see and use your unbiased opinion to make purchase decisions, we at Packt can understand what you think about our products, and our authors can see your feedback on their book. Thank you!

For more information about Packt, please visit packt.com.

Share Your Thoughts

Once you've read *Machine Learning on Kubernetes*, we'd love to hear your thoughts! Scan the QR code below to go straight to the Amazon review page for this book and share your feedback.

https://packt.link/r/1-803-24180-2

Your review is important to us and the tech community and will help us make sure we're delivering excellent quality content.

Part 1:
The Challenges of Adopting ML and Understanding MLOps
(What and Why)

In this section, we will define what MLOps is and why it is critical to the success of your AI journey. You will go through the challenges organizations may encounter in their AI journey and how MLOps can assist in overcoming those challenges.

The last chapter of this section will provide a refresher on Kubernetes and the role it plays in bringing MLOps to the OSS community. This is by no means a guide to Kubernetes, and you should consult other sources for a guide on Kubernetes.

This section comprises the following chapters:

- *Chapter 1, Challenges in Machine Learning*
- *Chapter 2, Understanding MLOps*
- *Chapter 3, Exploring Kubernetes*

1
Challenges in Machine Learning

Many people believe that **artificial intelligence** (**AI**) is all about the idea of a humanoid robot or an intelligent computer program that takes over humanity. The shocking news is that we are not even close to this. A better term for such incredible machines is human-like intelligence or **artificial general intelligence** (**AGI**).

So, what is AI? A more straightforward answer would be a system that uses a combination of data and algorithms to make predictions. AI practitioners call it **machine learning** or **ML**. A particular subset of ML algorithms, called **deep learning** (**DL**), refers to an ML algorithm that uses a series of steps, or layers, of computation (Goodfellow, Bengio, and Courville, 2017). This technique employs **deep neural networks** (**DNNs**) with multiple layers of artificial neurons that mimic the architecture of the human brain. Though it sounds complicated enough, it does not always mean that all DL systems will have a better performance compared to other AI algorithms or even a traditional programming approach.

ML is not always about DL. Sometimes, a basic statistical model may be a better fit for a problem you are trying to solve than a complex DNN. One of the challenges of implementing ML is about selecting the right approach. Moreover, delivering an ML project comes with other challenges, not only on the business and technology side but also in people and processes. These challenges are the primary reasons why most ML initiatives fail to deliver their expected value.

In this chapter, we will revisit a basic understanding of ML and understand the challenges in delivering ML projects that can lead to a project not delivering its promised value.

The following topics will be covered:

- Understanding ML

- Delivering ML value

- Choosing the right approach

- Facing the challenges of adopting ML

- An overview of the ML platform

Understanding ML

In traditional computer programming, a human programmer must write a clear set of instructions in order for a computer program to perform an operation or provide an answer to a question. In ML, however, a human (usually an ML engineer or data scientist) uses data and an algorithm to determine the best set of parameters for a model to yield answers or predictions that are usable. While traditional computer programs provide answers using exact logic (*Yes/No, Correct/Wrong*), ML algorithms involve fuzziness (*Yes/Maybe/No, 80% certain, Not sure, I do not know*, and so on).

In other words, ML is a technique for solving problems by using data along with an algorithm, statistical model, or a neural network, to infer or predict the desired answer to a question. Instead of explicitly writing instructions on how to solve a problem, we use a bunch of examples and let the algorithm figure out the best way (the best set of parameters) to solve the problem. ML is useful when it is impossible or extremely difficult to write a set of instructions to solve a problem. A typical example problem where ML shines is **computer vision** (**CV**). Though it is easy for any normal human to identify a cat, it is impossible or extremely difficult to manually write code to identify if a given image is of a cat or not. If you are a programmer, try thinking about how you would write this code without ML. This is a good mental exercise.

The following diagram illustrates where DL and ML sit in terms of AI:

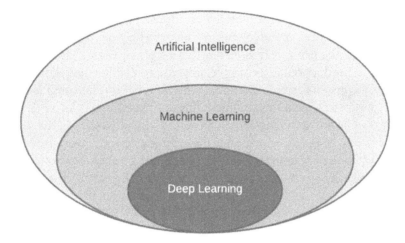

Figure 1.1 – Relationship between AI, ML, and DL

AI is a broad subject covering any basic, rule-based agent system that can replace a human operator, ML, and DL. But ML alone is another broad subject. It covers several algorithms, from basic linear regression to very deep **convolutional neural networks** (**CNNs**). In traditional programming, no matter which language or framework we use, the process of developing and building applications is the same. In contrast, ML has a wide variety of algorithms, and sometimes, they require a vastly different approach to utilize and build models from. For example, a **generative adversarial network** (**GAN**), which is an architecture used in many creative ML models to generate fake human faces, is trained differently to a basic decision tree model.

Because of the nature of ML projects, some practices in software engineering may not always apply to ML, and some practices, processes, and tools that are not present in traditional programming must be invented.

Delivering ML value

There are many books, videos, and lectures available on ML and its related topics. In this book, we will cover a more adaptive approach and show how **open source software** (**OSS**) can provide the basis for you and your organization to benefit from the AI revolution.

In later chapters, we will tackle the challenges behind operationalizing ML projects by deploying and using an open source toolchain on Kubernetes. Toward the end of the book, we will build a reusable ML platform that provides essential features that will help contribute to delivering a successful ML project.

Before we dig deeper into the software, we must have foundational knowledge, and we must know the practical steps required to successfully deliver business value with ML initiatives. With this knowledge, we will be able to address some of the challenges of implementing an ML platform and identify how they will help deliver the expected value from our ML projects. The primary reason why these promised values are not realized is that they don't get to production. For example, imagine you built an excellent ML model that predicts the outcome of football World Cup matches, but no one could use it during the tournament. As a result, even though the model is successful, it failed to deliver its expected business value. Most organization's AI and ML initiatives are in the same state. The data science or ML engineering team may have built a perfectly working ML model that could have helped the organization's business and/or its customers; however, these models do not usually get deployed to production. So, what are the challenges teams face that prevent them from putting their ML models into production?

Choosing the right approach

Before deciding to use ML for a given project, understand the problem first and assess if it can be solved by ML. Invest enough time in working with the right stakeholder to see what the expectations are. Some problems may be better suited to traditional approaches, such as when you have predefined business rules for a given system. It is faster and easier to code rules than is it to train a model, plus you do not need a huge amount of data.

While deciding whether to use ML or not, you can think in terms of whether pattern-based results will work for your problem. If you are building a system that reads data from the frequent-flyer database of an airline to find customers to which you want to send a promotion, a rule-based system may also give you good and acceptable results. An ML-based system may give you better matches for certain scenarios, but will the time spent on building this system be worth it?

The importance of data

The efficiency of your ML model depends on the quality and accuracy of the data, but unfortunately, data collection and processing activities do not get the attention they deserve, which proves costly in later stages of the project in terms of the model not being suitable enough for the given task.

> *"Everyone wants to do the model work, not the data work."*
>
> *– Data Cascades in High-Stakes AI, Sambasivan et al. (see the Further reading section)*

The paper cited here discusses this challenge. An interesting example quoted in the paper is of a team building a model to detect a particular pattern from patient scans, which works brilliantly with test data. However, the model failed in production because the scans being fed onto the model contained tiny dust particles, resulting in the inferior performance of the model. This example is a classic case of a team being focused on model building and not on how it will be used in the real world.

One thing that teams should put focus on is data validation and cleansing. Many times, data is often missing or is not correct—for example, a string field in a number column, different date formats in the same field, or the same **identifier** (**ID**) for different records if the records come from different systems. All this data anomaly may result in an inefficient model that will lead to inferior performance.

Once you've been through this process and come to the decision that yes, ML is the way to go… what next?

Facing the challenges of adopting ML

Organizations are eager to adopt ML to drive their business growth. In many projects, the teams become too focused on technical brilliance while not delivering the business value expected from the ML initiative. This can cause early failures that may result in reduced investment for future projects. These are the two main challenges that businesses are facing in making ML mainstream in all the various parts of the business, as outlined here:

- Keeping the focus on the big picture
- Siloed teams

Focusing on the big picture

The first challenge organizations face is building an ecosystem where ML models create value for the business. The challenging part is that teams often do not focus on all aspects of a project and instead focus only on specific areas, resulting in poor value for the business.

How many organizations that we know of are successful in their ML journey? Beyond the Googles, Metas (formerly Facebook), and Netflixs of the world, there are few success stories. The number one reason is that the teams put focus *just* on building the model. So, what else is there beyond the algorithm? Google published a paper about the hidden technical debt in ML projects (see the *Further reading* section at the end of this chapter), and it provides a good summary of things that we need to consider to be successful.

Have a look at the following diagram:

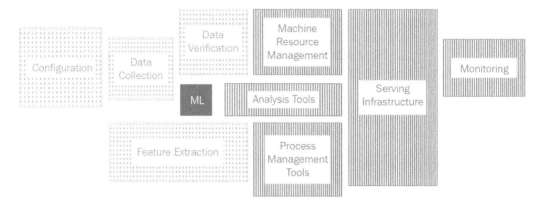

Figure 1.2 – The components of an ML system

Can you see the small block in *Figure 1.2*? The block in the picture captioned **ML** is the ML model development part, and you can see that there are a lot more processes involved in ML projects. Let's understand a few of them, as follows:

- **Data collection and data verification**: To have a reliable and trustworthy model, we need a good set of data. ML is all about finding patterns in the data and predicting unseen data results using those patterns. Therefore, the better the quality of your data, the better your model will perform. The data, however, comes in all shapes and sizes. Some of it may reside in files, some in proprietary databases; a dataset may come from data streams, and some data may need to be harvested from **Internet of Things** (**IoT**) devices. On top of that, the data may be owned by different teams with different security and regulatory requirements. Therefore, you need to think about technologies that allow you to collect, transform, and process data from various sources and in a variety of formats.

- **Feature extraction and analysis**: Often, assumptions about data quality and completeness are incorrect. Data science teams perform an activity called **exploratory data analysis** (**EDA**) in which they read and process data from various sources as fast as they can. Teams further improve their understanding of the data before they invest time in processing the data at scale and going to the model-building stage. Think about how your team or organization can facilitate the data exploration to speed up your ML journey.

Data analysis leads to a better understanding of data, but **feature extraction** is another thing. This is a process of identifying, through experiments, a set of data attributes that influences the accuracy of the model output and identifying which attributes are considered irrelevant or considered noise. For example, in an ML model that classifies if a bank transaction is fraudulent or not, the name of the account holder is considered to be irrelevant, or noise, while the amount of the transaction could be an important feature. The output of this process is a transformed version of the dataset that contains only relevant features and is formatted for consumption in the ML model training process or fitness function. This is sometimes called a **feature set**. Teams need a tool for performing such analysis and transforming data into a format that is consumable for model training. Data collection, feature extraction, and analysis are also collectively called **feature engineering** (**FE**).

- **Infrastructure, monitoring, and resource management**: You need computers to process and explore data, build and train your models, and deploy ML models for consumption. All these activities need processing power and storage capacity, at the lowest possible cost. Think about how your team will get access to hardware resources on-demand and in a self-service fashion. You need to plan how data scientists and engineers will be able to request the required resources in the fastest manner. At the same time, you still need to be able to follow your organization's policies and procedures. You also need system monitoring to optimize resource utilization and improve the operability of your ML platform.

- **Model development**: Once you have data available in the form of consumable features, you need to build your models. Model building requires many iterations with different algorithms and different parameters. Think about how to track the outcomes of different experiments and where to store your models. Often, different teams can reuse each other's work to increase the velocity of the teams further. Think about how teams can share their findings. Teams must have a tool that can facilitate model training and experiment runs, record model performance and experiment metadata, store models, and manage the tagging of models and promotion to an acceptable and deployable state.

- **Process management**: As you see, there are a lot of things to be done to make a useful model. Think about the processes of automating model deployment and monitoring processes. Different personas would be working on different things such as data tasks, model tasks, infrastructure tasks, and more. The team needs to collaborate and share to achieve a particular outcome. The real world keeps on changing: once your model is deployed into production, you may need to retrain your model with new data regularly. All these activities need well-defined processes and automated stages so that the team can continue working on high-value tasks.

In summary, you will need an ecosystem that can provide solution components for all of the following building blocks. This single platform will increase the team's velocity via consistent experience within the team for all the needs of an ML system:

- Fetching, storing, and processing data

- Training, tuning, and tracking models

- Deploying and monitoring models

- Automating repetitive tasks, such as data processing and model deployment

But how can we make different teams collaborate and use a common platform to do their tasks?

Breaking down silos

To complete an ML project, you need to have a team that comprises various roles. However, with diverse roles, there comes a challenge of communication, team dynamics, and conflicting priorities. In enterprises, these roles often belong to different teams in different **business units** (**BUs**).

ML projects need a variety of teams and personas to be successful. The following screenshot shows some of the roles and responsibilities that are required to complete a simple ML project:

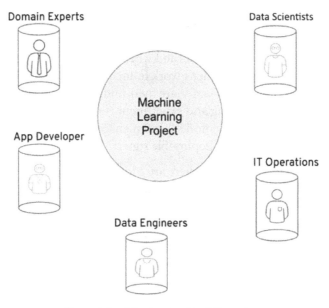

Figure 1.3 – Silos involved in ML projects

Let's look at these roles in more detail here:

- **Data scientist**: This role is the most understood one. This persona or team is responsible for exploring the data and running experiment iterations to determine which algorithm is suitable for a given problem.

- **Data engineers**: The persona or team in this role is responsible for ingesting data from various sources, cleaning the data, and making it useful for the data science teams.

- **Developers and operations**: Once the model is built, this team is responsible for taking the model and deploying it to be used. The operations team is responsible for making sure that computers and storage are available for the other teams to perform data processing, model life-cycle operations, and model inference.

- **A business subject-matter expert (SME)**: Even though data scientists build the ML model, understanding data and the business domain is critical to building the right model. Imagine a data scientist who is building a model for predicting COVID-19 without understanding the different parameters. An SME, which would be a medical doctor in this case, would be required to assist the data scientists in understanding data before going on to the model-building phase.

Of course, even with the building blocks in place, you're unlikely to succeed at the first attempt.

Fail-fast culture

Building a cross-functional team is not enough. Make sure that the team is empowered to make its own decisions and feels comfortable experimenting with different approaches. The data and ML fields are fast-moving, and the team may choose to adapt a recent technology or process or let go of an existing one based on the given success criteria.

Form a team of people who are passionate about the work, and when you give them autonomy, you will have the best possible outcome. Enable your teams so that they can adapt to change quickly and deliver value for your business. Establish an iterative and fast feedback cycle where teams receive feedback on work that has been delivered so far. A quick feedback loop will put more focus on solving the business problem.

However, this approach brings its own challenges. Adopting modern technologies may be difficult and time-consuming. Think of Amazon Marketplace: if you want to sell some new hot thing, by using Amazon Marketplace, you can bring your product to market faster because the marketplace takes care of a lot of moving parts required to make a sale. The ML platform you will learn about in this book enables you to experiment with modern approaches and modern technologies with ease by supplying basic common services and sandbox environments for your team to experiment fast.

It is critical to the success of projects that teams that belong to distinct groups form a cross-functional and autonomous team. This new team will move with higher velocity without internal friction and avoid tedious processes and delays. It is critical that the cross-functional team is empowered to drive its own decisions and be supported with self-serving platforms so it can work in an independent manner. The ML platform you will see in this book will provide the basis of one such platform where teams can collaborate and share.

Now, let's take a look at what kind of platform will help you address the challenges we have discussed.

An overview of the ML platform

In this section, we will talk about the capabilities of the ML platform that you will need to consider. The aim is to make you aware of the basic building blocks that could form an ecosystem for your team to help you in your ML journey. An ML platform can be thought of as a set of components that assists in the faster development and deployment of ML models and data pipelines.

There are three main characteristics of an ML platform, as outlined here:

- **A complete ecosystem**: The platform should provide an **end-to-end** (**E2E**) solution that includes data life-cycle management, ML life-cycle management, application life-cycle management, and observability.

- **Built on open standards**: The platform should provide a way to extend and build on the existing baseline. Because the field is fast-moving, it is critical that you can further enhance, tailor, and optimize platforms for your specific needs.

- **Self-serving**: The platform should be able to provide the resources required by teams automatically and on-demand, from hardware requests to deploying software in production. The platform automates the provisioning of resources based on enterprise controls and recovers them once the job is completed. The resources can be **central processing units** (**CPUs**), memory, or disk, or can be software such as **integrated development environments** (**IDEs**) to write code or a combination of these.

The following diagram shows the various components of an ML platform that serves different personas, allowing them to collaborate on a common platform:

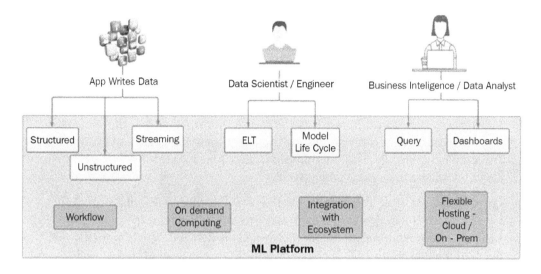

Figure 1.4 – Personas and their interaction with the platform

Apart from the characteristics presented in *Figure 1.4*, the platform must have the following technical capabilities:

- **Workflow automation**: The platform should have some form of workflow automation capability where both data engineers can create jobs that perform repetitive tasks such as data ingestion and preparation and data scientists can orchestrate model training and automate model deployments.

- **Security**: The platform must be secured to prevent data leaks and data loss that can have a negative impact on the business.

- **Observability**: We do not want to run applications without observability, whether it is a traditional application or an ML model. Deploying applications in production without observability is like riding a bike blindfolded. The platform should have a good amount of observability where you can monitor the health and performance of the entire system or sub-system in near real time. This should also include an alerting capability.

- **Logging**: Logging plays a key role in understanding what happened when systems start behaving in an unexpected way. The platform must have a solid logging mechanism to allow operations teams to better support the ML project.

- **Data processing and pipelining**: Because ML projects rely on a huge amount of data, the platform must include a reliable fully featured data processing and data pipelining solution that can scale horizontally.

- **Model packaging and deployment**: Not all data scientists are experienced software engineers. Although some may have an experience in writing applications, it is not safe to assume that all data scientists can write production-grade applications and deploy them to production. Therefore, the platform must be able to automatically package an ML model into an application and serve it.

- **ML life cycle**: The platform must also be capable of managing ML experiments, tracking performance, storing training and experiment metadata and feature sets, and versioning models. This not only allows data scientists to work efficiently, but also allows them to work collaboratively.

- **On-demand resource allocation**: One important feature an ML platform should have is the capability that allows data scientists and data engineers to provision their own runtime resources automatically and on-demand. This eliminates the need for manual requisition of resources and eliminates time wasted on waiting and handovers with operations teams. The platform must allow platform users to create their own environment and to allocate the right amount of compute resources they need to do their jobs.

There are already platform products that have most, if not all, of the capabilities you have just learned about. What you will learn in the later chapters of this book is how to build one such platform based on OSS on top of Kubernetes.

Summary

Even though ML is not new, recent advancements in relatively cheap computing power have allowed many companies to start investing in it. This widespread availability of hardware comes with its own challenges. Often, teams do not put the focus on the big picture, and that may result in ML initiatives not delivering the value they promise.

In this chapter, we have discussed two common challenges that enterprises face while going through their ML journey. The challenges span from the technology adoption to the teams and how they collaborate. Being successful with your ML journey will require time, effort, and practice. Expect it to be more than just a technology change. It will require changing and improving the way you collaborate and use technology. Make your team autonomous and prepare it to adapt to changes, enable a fail-fast culture, invest in technology, and always keep an eye on the business outcome.

We have also discussed some of the important attributes of an E2E ML platform. We will talk about this topic in-depth in the later parts of this book.

In the next chapter, we will introduce an emerging concept in ML projects, **ML operations** (**MLOps**). Through this, the industry is trying to bring the benefits of software engineering practices to ML projects. Let's dig in.

Further reading

If you want to learn more about the challenges in machine learning, you might be interested in the following articles as well.

- *Hidden Technical Debt in Machine Learning, Sculley et al.*, 2015: `https://papers.nips.cc/paper/5656-hidden-technical-debt-in-machine-learning-systems.pdf`

- *Data Cascades in High-Stakes AI, Sambasivan et al.*, 2021: `https://storage.googleapis.com/pub-tools-public-publication-data/pdf/0d556e45afc54afeb2eb6b51a9bc1827b9961ff4.pdf`

2
Understanding MLOps

Most people from software engineering backgrounds know about the term **development-operations** (**DevOps**). To us, DevOps is about collaboration and shared responsibilities across different teams during the **software development life cycle** (**SDLC**). The teams are not limited to a few **information technology** (**IT**) teams; instead, it involves everyone from the organization who is a stakeholder in the project. No more segregation between building software (developers' responsibility) and running it in production (operations' responsibility). Instead, the team *owns* the product. DevOps is popular because it helps teams increase the velocity and reliability of the software being developed.

In this chapter, we will cover the following topics:

- Comparing **machine learning** (**ML**) to traditional programming
- Exploring the benefits of DevOps
- Understanding **ML operations** (**MLOps**)
- The role of **open source software** (**OSS**) in ML projects
- Running ML projects on Kubernetes

Before we can apply DevOps to ML projects, we must first understand the difference between traditional software development and ML development processes.

Comparing ML to traditional programming

As with traditional application development, an ML project is also a software project, but there are fundamental differences in the way they are delivered. Let's understand how an ML project is different from a traditional software application.

In traditional software applications, a software developer writes a program that holds an explicitly handcrafted set of rules. At runtime or prediction time, the built software applies these well-defined rules to the given data, and the output of the program is the result calculated based on coded rules.

The following diagram shows the **inputs and outputs** (**I/Os**) for a traditional software application:

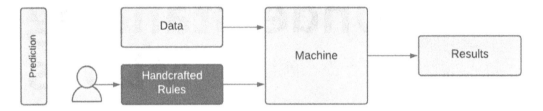

Figure 2.1 – Traditional software development

In an ML project, the rules or patterns are not completely known, therefore we cannot explicitly describe rules in code as we can in traditional programming. In ML, there is a process that extracts rules based on a given sample pair of data and its associated expected results. This process is called **model training**. In the model-training process, the chosen ML algorithm calculates rules based on the given data and the verified answer. The output of this process is the **ML model**. This generated model can then be used to infer answers during prediction time. In contrast with traditional software development, instead of using explicitly written rules, we use a generated ML model to get a result.

The following diagram shows that the ML model is generated at training time, which is then used to produce answers or results during prediction time:

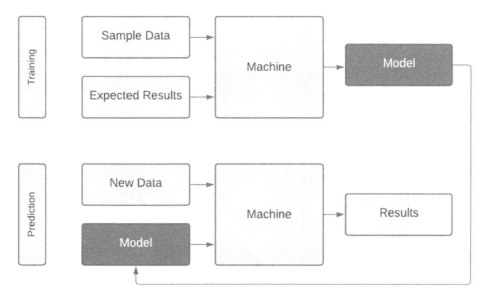

Figure 2.2 – ML development

Though traditional software development and ML are fundamentally different, there are some synergies in the engineering processes between the two approaches. Given that traditional software development is very mature in the current era, we can apply lessons from it to our ML projects. Primarily, of course, both traditional programming and ML are software. Whichever processes we apply to build software in the traditional world—such as versioning, packaging of software as containers, automated deployments, and so on—these can be applied to ML projects too. However, we also must accommodate added processes in ML, such as model training.

So, why do we really need DevOps in ML projects? What does it bring to the table? Let's have a look at this in the next section.

Exploring the benefits of DevOps

DevOps is not just about toolsets. Say you have a tool available that can run unit tests for you. However, if the team has no culture of writing test cases, the tool would not be useful. DevOps is about how we work together on tasks that span across different teams. So, the three primary areas to focus on in DevOps are these:

- **People**: Teams from multiple disciplines to achieve a common goal
- **Processes**: The way teams work together
- **Technology**: The tools that facilitate collaboration across different teams

DevOps is built on top of Agile development practices with the objective of streamlining the software development process. DevOps teams are cross-functional, and they have the autonomy to build software through **continuous integration/continuous delivery (CI/CD)**. DevOps encourages teams to collaborate over a fast feedback loop to improve the efficiency and quality of the software being developed.

The following diagram illustrates a complete DevOps cycle for traditional software development projects:

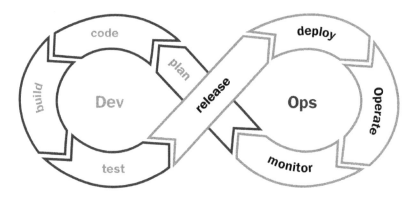

Figure 2.3 – A mobius loop showcasing a DevOps process

Through DevOps, teams can have well-defined and streamlined development practices for building, testing, deploying, and monitoring software in production. All this makes it possible to quickly and reliably release software into production. Some of the benefits that come out of DevOps practices are presented here:

- **CI/CD**: CI is a phase through which software is merged and verified as soon as the developer pushes it into the code repository. CD is a series of stages through which software is built, tested, and packaged in a deployment ready form. **Continuous deployment** (also known as **CD**) is a phase where the deployment-ready code is picked and deployed to be consumed by end users. In DevOps, all these processes are automated.

- **Infrastructure as Code (IaC)**: IaC is an approach to automate the provisioning and configuring of IT infrastructure. This aspect enables the team to request and configure infrastructure on an on-demand and as-needed basis. Imagine that a data scientist in your team needs a **graphics processing unit (GPU)** to do their model training. If we follow the practice of configuring and provisioning IaC, the request for a GPU can be automatically fulfilled by the system. In the next chapters, you will see this capability in action.

- **Observability**: Observability relates to how well we understand the state of our running system. DevOps makes systems observable via federating logging from different components, monitoring the systems (such as **central processing unit** (**CPU**), memory, response times, and so on), and providing a way to correlate various parts of the system for a given call through call tracing. All these capabilities, collectively, provide the basis for understanding the system state and help debug any issues without changing the code.

- **Team collaboration**: DevOps is not just about technology. In fact, the key focus area for the team is to collaborate. Collaboration is how multiple individuals from different teams work toward a common goal. Business, development, and operations teams working together is the core of DevOps. For ML-based projects, the team will have data scientists and data engineers on top of the aforementioned roles. With such a diverse team, communication is critical for building collective understanding and ownership of the defined outcome.

So, how can we bring the benefits of a DevOps approach to ML projects? The answer is MLOps.

Understanding MLOps

MLOps is an emerging domain that takes advantage of the maturity of existing software development processes—in other words, DevOps combined with data engineering and ML disciplines. MLOps can be simplified as an engineering practice of applying DevOps to ML projects. Let's take a closer look at how these disciplines form the foundation of MLOps.

ML

ML projects involve activities that are not present in traditional programming. You learned in *Figure 2.3* that the bulk of the work in ML projects is not model development. Rather, it is more data gathering and processing, data analysis, **feature engineering** (**FE**), process management, data analysis, model serving, and more. In fact, according to the paper *Hidden Technical Debt in Machine Learning Systems* by D. Sculley et al., only 5% of the work is ML model development. Because of this, MLOps is not only focused on the ML model development task but mostly on the big picture—the entire ML project life cycle.

Just as with DevOps, MLOps focuses on people, processes, and technology. But there are some complexities that MLOps has to address and DevOps doesn't have to. Let's look at some of these complexities in more detail here:

- First, unlike traditional programming, where your only input is code, in ML, your input is both code and data. The ML model that is produced in the model development stage is highly dependent on data. This means that even if you do not change your code, if you train an ML algorithm using a different dataset, the resulting ML model will be different and will perform differently. When it comes to **version control**, this means that you not only version the code that facilitates model training, but you also need to version the data. Data is difficult to version because of the huge amount required, unlike code. One approach to address this is by using Git to keep track of a dataset version using the hash of the data. The actual data is then stored somewhere in remote storage such as a **Simple Storage Service (S3)** bucket. An open source tool called **Data Version Control (DVC)** can do this.

- Secondly, there are more personas involved and *more collaboration* required in ML projects. You have data scientists, ML engineers, and data engineers collaborating with software engineers, business analysts, and operations teams. Sometimes, these personas are very diverse. A data scientist may not completely understand what production deployment really is. On the other hand, operations people (and sometimes even software engineers) do not understand what an ML model is. This makes collaboration in ML projects more complicated than a traditional software project.

- Third, the addition of a model development stage adds more pivot points to the life cycle. This complicates the whole process. Unlike traditional software development, you only need to develop one set of working code. In ML, a data scientist or ML engineer may use multiple ML algorithms and generate multiple resulting ML models, and because only one model will get selected to be deployed to production, those models are compared with each other in terms of performance against other model properties. MLOps accommodates this complex workflow of *testing, comparing, and selecting models* to be deployed to production.

Building traditional code to generate an executable binary usually takes a few seconds to a few minutes. However, *training an ML algorithm to produce an ML model can take hours or days*, sometimes even weeks when you use certain **deep learning (DL)** algorithms. This makes setting up an Agile iterative time-bound cadence a little complicated. An MLOps-enabled team needs to handle this delay in their workflow, and one way to do this is to start building the other model while waiting for other models to be trained completely. This is very difficult to achieve if the data scientists or ML engineers are training their ML algorithms using their own laptops. This is where the use of a scalable infrastructure comes in handy.

- Lastly, because ML models' performances rely on the data used during training, if this data no longer represents the real-world situation, the model accuracy will degrade, resulting in poor prediction performance. This is called **model drift**, and this needs to be detected early. This is usually incorporated as part of the monitoring process of the ML project life cycle. Aside from the traditional metrics that you collect in production, with ML models, you also need to monitor model drift and outliers. Outlier detection, however, is much more difficult to implement, and sometimes requires you to train and build another ML model. **Outlier detection** is about detecting incoming data, in production, that does not look like the data the model was trained on: you do not want your model to provide irrelevant answers to these non-related questions. Another reason is that this could be an attack or an attempt to abuse the system. Once you have detected model drift or outliers, what are you going to do with this information? It could very well be just about raising an alert, or it could trigger some other automated processes.

Because of the complexity ML adds when compared to traditional programming, the need to address these complexities led to the emergence of MLOps.

DevOps

In terms of deployment, think about all the sets of code you write in an ML project: the code that performs the data processing, the code that facilitates model training and FE, the code that runs the model inference, and the code that performs model drift and outlier detection. All of these sets of code need to be built, packaged, and deployed for consumption at scale. This code, once running in production, needs to be monitored and maintained as well. This is where the CI/CD practices of DevOps help. The practice of automating software packaging, testing, securing, deploying, and monitoring came from DevOps.

Data engineering

Every ML project involves **data engineering**, and ML projects deal with a lot of data a lot more than code. Therefore, it is mandatory that your infrastructure includes data processing capabilities and that it can integrate with existing data engineering pipelines in your organization.

Data engineering is a huge subject—an entire book could be written about it. But what we want to emphasize here is that MLOps intersects with data engineering practices, particularly in **data ingestion**, **data cleansing**, **data transformation**, and **big data testing**. In fact, your ML project could be just a small ML classification model that is a subpart of a much bigger data engineering or data analytics project. MLOps adopts the best practices in data engineering and analytics.

A representation of MLOps is provided in the following diagram:

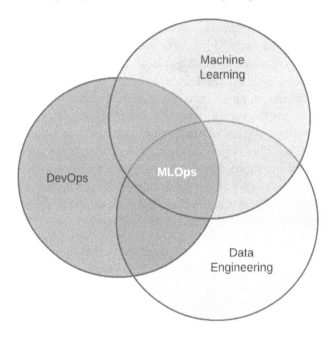

Figure 2.4 – MLOps as the intersection of ML, data engineering, and DevOps

To put it in another way, MLOps, as shown in *Figure 2.4*, is the convergence of **ML**, **DevOps**, and **data engineering** disciplines that focus on running ML in production. It is also about encapsulating ML projects in a highly scalable, reliable, observable infrastructure. Finally, it is also about establishing repeatable processes for teams to perform the tasks required to successfully deliver ML projects, as shown in *Figure 2.4*, while supporting collaboration with each other.

With this basic understanding of MLOps, let's dig a little deeper into the ML project life cycle. We'll start by defining what are the general stages of an ML project.

ML project life cycle

As with DevOps, which provides a series of activities that could be performed in a DevOps cycle, you can see a series of steps that could be used to take your ML project from start to finish in *Figure 2.5*. These steps or stages will become part of your ML projects' life cycle and provide a consistent way to take your ML projects into production. The ML platform that you build in this book is the ecosystem that allows you to implement this flow. In later chapters of this book, you will use this flow as the basis for the platform. A summary of the stages in an ML project could be depicted as follows:

Figure 2.5 – A ML project life cycle

Here is a definition of each stage of the project life cycle presented in the preceding diagram:

- **Codify the problem and define success metrics**: In this stage, the team evaluates if the given business problem can be solved using ML. Notice the word *team* here, which would consist of data scientists and the business **subject-matter expert (SME)** at a minimum. The team will then define a success criterion to assess the prediction of the model.

- **Ingest, clean, and label data**: In this stage, the team assesses if the data required to train the model is available. The team will play an additional role, that of data engineers, to help move the project during this stage and beyond. The team will build components to ingest data from a variety of sources, clean the captured data, possibly label the data, and store it. This data will form the basis of ML activities.

- **FE**: FE is about transforming the raw data into features that are more relevant to the given problem. Consider you are building a model that predicts if any given passenger on the *Titanic* will survive or not. Imagine the dataset you got contains the ticket number of the passenger. Do you think ticket numbers have something to do with the survival of the passenger? A business SME may mention that ticket numbers may be able to provide which class the customer belongs to on the ship, and first-class passengers may have easier access to lifeboats on the ship.

- **Model building and tuning**: In this stage, the team starts experimenting with different models and different hyperparameters. The team will test the model against the given dataset and compare the results of each iteration. The team will then determine the best model for the given success metrics and store the model in the model registry.

- **Model validation**: In this stage, the team validates the model against a new set of data that is not available at the training time. This stage is critical as it **determines** if the model is generalized enough for the unseen data, or if the model only works well on the training data but not on the unseen data—in other words, avoiding **overfitting**. Model validation also involves identifying **model biases**.

- **Model deployment**: In this stage, the team picks the model from the model registry, packages it, and deploys it to be consumed. Traditional DevOps processes could be used here to make the model available as a service. In this book, we will focus on **model as a service** (**MaaS**), where the model is available as a **REpresentational State Transfer** (**REST**) service. However, in certain scenarios, the model could be packaged as a library for other applications to use it.

- **Monitoring and validation**: In this stage, the model will be continually monitored for response times, the accuracy of predictions, and whether the input data is like the data on which the model is trained. We have briefly touched on outlier detection. In practice, it works like this: imagine that you have trained your model for rush-hour vacancy in a public transport system, and the data the model is trained against is where citizens use the public transport system for over a year. The data will have variances for weekends, public holidays, and any other events. Now, imagine if, due to the COVID-19 lockdown, no one is allowed to use the public transport system. The real world is not the *same* as compared to the data our model is trained upon. Naturally, our model is not particularly useful for this changed world. We will need to detect this anomaly and generate alerts so that we can retrain our model with the new datasets if possible.

You have just learned the stages of the ML project life cycle. Although the stages may look straightforward, in the real world, there are several good reasons why you need to go back to previous stages in certain cases.

Fast feedback loop

A keen observer may have noticed that a key attribute of the Agile and cross-functional teams that we presented in the first chapter is not available in the stages presented so far in this chapter. Modern DevOps is all about fast feedback loops to course-correct early in the project life cycle. The same concept will bring even more value to ML projects because ML projects are more complex than traditional software applications.

Let's see at which stages we can assess and evaluate the progress of the team. After evaluation, the team can decide to course-correct by going back to an earlier stage or moving on to the next stage.

The following diagram shows the ML project life cycle with feedback checkpoints from various stages, denoted by green arrows:

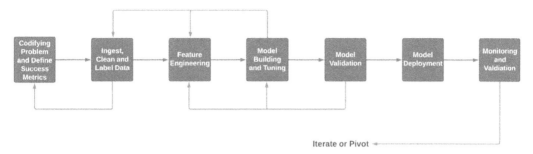

Figure 2.6 – A ML project life cycle with feedback checkpoints

Let's look at this in more detail here:

- **Checkpoint from the ingest, clean, and label data stage**: After *Stage 1* is completed, you have started to process data as defined in the second stage. You may find that the actual data is incomplete or not correct. You can take this feedback to improve your understanding of data and may need to redefine the success criteria of the project or, in worse cases, stop the project because the required data is not available. In many scenarios, teams find additional data sources to fill the data gap identified in the second stage.

- **Checkpoint from the model building and tuning stage**: During this stage, the team may find that the features available to train the model may not be enough to get the desired metric. At this point, the team may decide to invest more time in finding new features or revisit the raw data to determine if more data is needed.

- **Checkpoint from the model validation stage**: During this stage, the model will be validated against a new dataset that the model has never seen before. Poor metrics at this stage may trigger the tuning of the model, or you may decide to go back to find more features for better model performance,

- **Checkpoint from the model monitoring and validation stage**: Once the model moves into production, it must be monitored continuously to validate if the model is still relevant to the real and changing world. You need to find out if the model is still relevant and, if not, how you can make the model more useful. The result of this may trigger any other stage in the life cycle; as you can see in *Figure 2.6*, you may end up retraining an existing model with new data or going to a different model altogether, or even rethinking if this problem *should* be tackled by ML. There is no definitive answer on which stage you end up at; just as with the real world, it is unpredictable. However, what is important is the capability to re-assess and re-evaluate, and to continue to deliver value to the business.

You have seen the stages of the ML project life cycle and the feedback checkpoints from which you decided whether to continue to the next stage or go back to previous stages. Now, let's look at the personas involved in each of the stages and their collaboration points.

Collaborating over the project life cycle

We have defined a streamlined process for building our model. Let's try to define how a team of diverse roles and abilities will collaborate on this model. Recall from the previous chapter that building a model takes effort from different teams with different abilities. It is important to note that in smaller projects, the same person may be representing different roles at the same time. For example, in a small project, the same person can be both a data scientist and a data engineer.

The following diagram shows an ML project life cycle with an overlay of feedback points and personas:

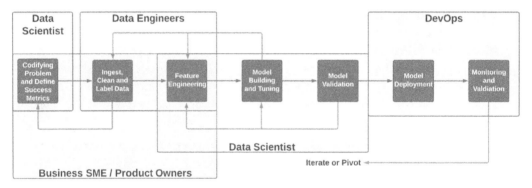

Figure 2.7 – A ML project life cycle with feedback checkpoints and team roles

The ML project within your organization needs collaboration between data scientists and the business SMEs in the first stage. Imagine the team wants to predict, based on a picture, the probability of a certain type of skin disease.

- At this stage, a collaboration between data scientists and doctors (the SME for this case) is needed to define the problem and the performance metrics. Without this collaboration, the project would not be successful.

- In the second stage—the data ingestion and cleaning stage—data engineers will need to work along with the business SMEs to understand which data is available and how to clean and label it correctly. The knowledge the SMEs will bring during this stage is critical as this is responsible for creating a useful dataset for future stages.

- In the third stage, data scientists, data engineers, and SMEs will collaborate to work on the base data from the second stage and process it to extract useful features from it. The data scientists and SMEs will provide guidance on which data can be extracted, and the data engineer will write processing logic to do so.

- In the fourth and the fifth stages, most of the work will be done by data scientists to build and tune the model as per the given criteria. However, based on whether or not the model has managed to achieve the defined metric, the team may decide to go back to any of the previous stages for better performance.

Once the model is built, the DevOps team experts can package, version, and deploy the model to the correct environment.

- The last stage is critical: the team uses observability capabilities to monitor the performance of the model in the production environment. After monitoring the model performance in the real world and based on the feedback, the team may again decide to go back to any of the previous stages to make the model more useful for the business.

Now that you have a good understanding of the challenges we have highlighted and how you can overcome these challenges using the ML life cycle, the next phase is to have a platform that supports this life cycle while providing a solution for each component defined in the big picture (see *Chapter 1*, *Challenges in Machine Learning*) with self-service and automation capabilities. What better way to start a journey while collaborating with the open source community?

The role of OSS in ML projects

Now that you have a clear understanding of what problems the ML platform is expected to solve, let's see why open source is the best place to start. We should start with some definitions to set the basics, right?

Free OSS is where *the users have the freedom to run, copy, distribute, study, change, and improve the software.*

> **OSS**
> For more information on OSS, see the following link:
> `https://www.gnu.org/philosophy/free-sw.html`

OSS is everywhere. Linux is the most common operating system, running in data centers and powering the cloud around the world. Apache Spark and related open source technologies are the foundation for the big data revolution for a range of organizations. Open source-based **artificial intelligence** (**AI**) technologies such as TensorFlow and MLflow are at the forefront of AI advancement and are used by hundreds of organizations. Kubernetes, the open source container orchestration platform, has become the de facto standard for container platforms.

The top players in computing—such as Amazon, Apple, Facebook, Google, Microsoft, and Red Hat, to name a few—have contributed to and owned major open source projects, and fresh players are joining all the time. Businesses and governments around the world depend on open source to power mission-critical and highly scalable systems every day.

One of the most successful open source projects in the cloud computing space is **Kubernetes**. Kubernetes was founded in mid-2014 and was followed by the release of its version 1.0 in mid-2015. Since then, it has become the de facto standard for container orchestration.

Moreover, the **Cloud Native Computing Foundation** (**CNCF**) was created by *The Linux Foundation* with the mission of making cloud computing ubiquitous. CNCF does this by bringing together the world's top engineers, developers, end users, and vendors. They also run the world's largest open source conferences. The foundation was created by using **Kubernetes** as the seed project. This is how Kubernetes sets the standard definition of **cloud native**. As of this writing, the foundation has 741 member organizations and 130 certified Kubernetes distributions and platforms and has graduated 16 very successful open source projects. Among those projects is, of course, **Kubernetes** but also the **Operator Framework**, which you will learn more about in the next chapter.

Before the explosion of **big data** and **cloud computing**, ML projects were mostly academic. They seldom left the boundaries of colleges and universities, but this doesn't mean that AI, ML, and **data science** were not progressing forward. The academic world has actually created hundreds of open source Python libraries for mathematical, scientific, and statistical calculations. These libraries have become the foundation modern ML frameworks are built upon. The most popular ML frameworks at the time of writing—TensorFlow, PyTorch, scikit-learn, and Spark ML—are all open source. The most popular data science and ML development environments today—Jupyter Notebook, JupyterLab, JupyterHub, Anaconda, and many more—are also all open source.

ML is an evolving field, and it needs the vision of larger communities that go beyond any single organization. The process of working in a community-based style enables the collaboration and creativity that is required by ML projects, and open source is an important part of why ML is progressing at a tremendous speed.

You now have a basic understanding of how important OSS is in the AI and ML space. Now, let's take a closer look at why you should run ML projects on Kubernetes.

Running ML projects on Kubernetes

For building reliable and scalable ML systems, you need a rock-solid base. **Kubernetes** provides the foundation for building scalable and reliable distributed systems along with the self-service capabilities that are required by our platform. The capability of Kubernetes to abstract the hardware infrastructure and consume it as a single unit is of great benefit to our platform.

Another key component is the ability of Kubernetes-based software to run anywhere, from small on-premises data centers to large hyperscalers (**Amazon Web Services** (**AWS**), **Google Cloud Platform** (**GCP**), Azure). This capability will give you the portability to run your ML platform anywhere you want. The consistency it brings to the consumer of your platform is brilliant as the team can experiment with extremely low initial costs on the cloud and then customize the platform for a wider audience in your enterprise.

The third and final reason to opt for Kubernetes is its capability to run different kinds of workloads. You probably remember from the previous chapter that a successful ML project needs not only ML but also infrastructure automation, data life cycle management, stateful components, and more. Kubernetes provides a consistent base to run diverse types of software components to create an **end-to-end** (**E2E**) solution for business use cases.

The following screenshot shows the layers of an ML platform. Kubernetes provides the scaling and abstracting layer on which an ML platform is built. Kubernetes offers the freedom of abstracting the underlying infrastructure. Because of this flexibility, we can run on a variety of cloud providers and on-premises solutions. The ML platform you will build in this book allows operationalization and self-service in the three wider areas of an ML project—FE, model development, and DevOps:

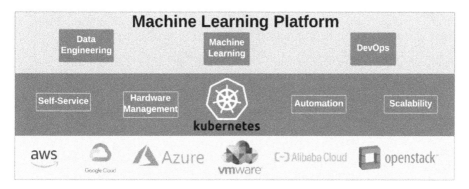

Figure 2.8 – An OSS-based ML platform

There you go: your ML platform will be based on OSS and will use Kubernetes as the hosting base. The strength of the open source Kubernetes communities will help you use the best technologies that will evolve as the field continues to mature.

Summary

In this chapter, we have defined the term *MLOps* and suggested an ML project life cycle that is collaborative and provides early feedback. You have learned that with this project life cycle, the team can continuously deliver value to the business. You have also learned about some of the reasons why building a platform based on OSS makes sense and the benefits of community-driven software.

This completes the part of the book about setting the context, learning why a platform is needed, and discovering what kinds of problems it is expected to solve. In the next chapter, we will examine some basic concepts of the Kubernetes system that is at the heart of our ML platform.

Further reading

For more information regarding the topics that were covered in this chapter, take a look at the following resources:

- *DevOps: Breaking the development-operations barrier* `https://www.atlassian.com/devops`

3
Exploring Kubernetes

Now that you have seen that Kubernetes will form the basis of your **machine learning (ML)** platform, it's logical to refresh your knowledge of the underlying bedrock of our solution. Though there are many resources available on the internet on this topic of Kubernetes, we will briefly discuss the role of Kubernetes in the cloud era and the flexibility it provides for building solutions. You will also learn about Operators in Kubernetes and how they help simplify the installation and operation of Kubernetes workloads. By the end of this chapter, you will have built a running `minikube` instance either in your local machine or in the cloud. This is a single-node Kubernetes cluster that you will use as the base infrastructure to build and run the ML platform.

In this particular order, we will cover the following topics:

- Exploring Kubernetes major components
- Becoming cloud-agnostic through Kubernetes
- Understanding Operators
- Setting up your local Kubernetes environment
- (Optional) Provisioning a **virtual machine** (**VM**) on **google cloud platform** (**GCP**)

Technical requirements

This chapter includes some hands-on setup. You will be setting up a Kubernetes cluster, and for this, you will need a machine with the following hardware specifications:

- A **central processing unit** (**CPU**) with at least four cores; eight are recommended
- Memory of at least 16 **gigabytes** (**GB**); 32 GB is recommended
- Disk with available space of at least 60 GB

This can be a physical machine such as a laptop, a server, or a VM running in the cloud that supports nested virtualization.

Exploring Kubernetes major components

There are many definitions of Kubernetes available on the web. We assume that, as a Kubernetes user, you already have a favorite pick. Therefore, in this section, you will see some basic concepts to refresh your Kubernetes knowledge. This section is by no means a reference or tutorial for the Kubernetes system.

From *Chapter 2, Understanding MLOps*, you have seen that Kubernetes provides the means for your ML platform to perform the following capabilities:

- **Provide a declarative style of running software components**: This capability will help your teams to be autonomous.
- **Provide an abstraction layer for hardware resources**: Through this capability, you can run your ML platform on a variety of hardware and provide on-demand resource scheduling.
- **Provide an application programming interface (API) to interact with it**: This will enable you to bring the automation for running different components onto your ML platform.

Let's start by defining the major components of the Kubernetes platform: the control plane and the worker nodes.

Control plane

The **control plane** is a set of components that form the *brains* of the Kubernetes. It consists of an API server, a key-value database, a scheduler, and a set of controllers. Let's define each of these components, as follows:

- **API server**: This component provides a set of **REpresentational State Transfer** (**REST**) APIs to interact with the Kubernetes system. Everyone interacts with Kubernetes through this API. As a developer or operations engineer, you use the API, and internal Kubernetes components talk to the API server to perform different activities.

- **Key-value database**: The API server is stateless; it needs to have a persistent store where it can store different objects. The key-value database is fulfilled by a component called etcd. No other component of the Kubernetes system talks to this value store directly—this is only accessible by the API server.

- **Scheduler**: The scheduler component dictates where an application instance would be running. The scheduler selects the most suitable worker node based on the policy defined by the Kubernetes administrator.

- **Controllers**: There are multiple controllers running in the control plane. Each controller has a set task; for example, a node controller is responsible for monitoring the state of the nodes.

The following diagram shows the interaction between multiple control-plane components:

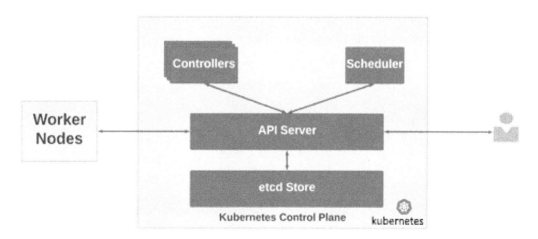

Figure 3.1 – Kubernetes control-plane components

The control plane orchestrates the creation, update, and deletion of objects. It monitors and maintains the healthy state of the Kubernetes cluster. The control plane runs workloads that keep the cluster running. But what about the application workloads?

Worker nodes

As the name suggests, workers are a set of nodes that host the application software. For example, all ML platform components will be executed on the worker nodes. However, worker nodes also run a couple of Kubernetes components that make the communication channel between the control plane and the worker and manage running applications on the worker node. These are the key components running on the worker nodes besides the applications:

- **Kube proxy**: Its primary role is to manage network communications rules for your applications running on the node.

- **Kubelet**: Think of the Kubelet software component as an agent running on each node. The primary role of this agent is to talk to the control-plane API server and manage applications running on the node. The agent also captures and sends the status of the node and the applications back to the control plane via the API.

- **Container runtime**: The container runtime component is responsible for running containers that host applications, as directed by the Kubelet. *Docker* is one such example; however, Kubernetes has defined a **container runtime interface** (**CRI**). CRI defines interfaces that Kubernetes uses and the Kubernetes administrator can choose any container runtime that is compatible with the CRI.

The following diagram shows the interaction between multiple worker-node components:

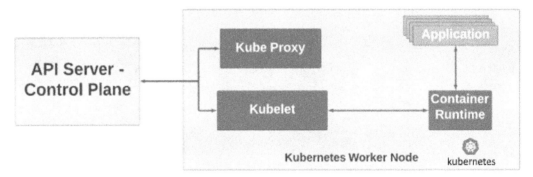

Figure 3.2 – Kubernetes worker components

Worker nodes, also known as compute nodes, do the actual work of running the application workloads in the cluster. Running application workloads requires you to interact with the control plane using Kubernetes objects or resources.

Kubernetes objects required to run an application

Now, let's define a set of Kubernetes **objects** that are commonly required to run an application on the Kubernetes system. When you build the components for your ML platform, you will be using these Kubernetes objects to run applications on top of Kubernetes. The objects are listed here:

- **Namespace**: One Kubernetes cluster is shared by multiple teams and projects. Namespaces provide a way to isolate Kubernetes resources. This isolation allows different teams, different environments, or even different applications to share the same cluster while keeping different configurations, network policies, resource quotas, and access control. It is like having a logical sub-cluster within the same Kubernetes cluster.

- **Container image**: When you want to run an application on Kubernetes, you need to package the application in a standard format. This packaged format, which consists of your application and all its dependencies, is called a container image, and the running instance of this image is called a **container**. It contains your application and all the dependencies, including the operating system resources and your application needs, in one common bundle.

- **Deployment**: This Kubernetes object represents an application's desired state on the cluster. A Deployment object contains information such as which container image you want to run and how many instances or *replicas* of containers you require. Kubernetes is periodically comparing the current cluster state to the desired state defined in the Deployment object. When Kubernetes finds that the current state is different from the desired state, it will then apply the necessary updates to the cluster to achieve the desired state. These updates include spinning up new containers with the container image defined in the Deployment object, stopping containers, and configuring network and other resources required by the Deployment object.

- **Pods**: A Pod is a fundamental unit of running applications in Kubernetes. It is also the smallest schedulable unit of deployment. It can contain one or more containers. Containers inside the pod share networking and disk resources. Containers running in a single pod are scheduled together on the same node while also having local communication with each other.

- **Services**: How do pods communicate with each other? Pods communicate through the cluster network, and each pod has its own **Internet Protocol** (**IP**) address. However, pods may come and go. Kubernetes may restart a pod due to node health or scheduling changes, and when this happens, the pod's IP address will change. Furthermore, if the Deployment object is configured to run multiple replicas of the same pods, this means each replica will have its own IP address.

  A service, in Kubernetes, exposes a set of pods as a single abstracted network service. It provides a single consistent IP address and **Domain Name System** (**DNS**) name that can route traffic and perform load balancing on pods. Think of a service as a load-balanced reverse proxy to your running pods.

- **ConfigMaps and Secrets**: We have our application packaged as a container image and running as a pod. The same pod will get deployed in multiple environments such as Dev, Test, and Production. However, each environment will have a different configuration, such as the database location or others. Hardcoding such a configuration into a container image is not the right approach. One reason is that the container image may be deployed in multiple environments with different environment settings. There must be a way to define configuration outside of the container image and inject this configuration onto our container at runtime!

  ConfigMaps and Secrets provide a way to store configuration data in Kubernetes. Once you have these objects defined, they can be injected into your running pods either as a file within the pod's filesystem or as a set of environment variables.

  A ConfigMap is used to store and access configuration data. However, for sensitive configurations such as passwords and private keys, Kubernetes provides a special object for this purpose, known as a Secret. Just as with ConfigMaps, Secrets can be mounted either as files or as environment variables into pods.

The following diagram shows a logical relationship between Deployments, pods, ConfigMaps, and Services. A Deployment object provides an abstraction of a containerized application. This hides the complexity behind running **replication controllers** and pods. Deployments help you in running your application as a pod or group of pods, ConfigMaps provide an environment-specific configuration for your pods, and Services expose the pods in your deployment as a single network service:

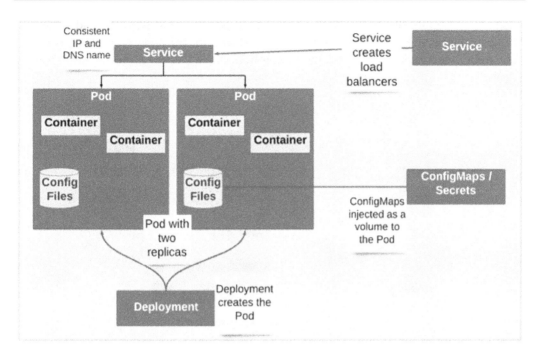

Figure 3.3 – Storage provisioning in Kubernetes

- **Storage (PersistentVolume and PersistentVolumeClaim (PV and PVC))**: Pods
are ephemeral. Once they are destroyed, all the local resources of the pod are gone.
More often, applications deployed as pods may need access to storage to read and
write persistent data that can outlive pods.

Kubernetes promises to be the infrastructure abstraction layer on top of many
hardware vendors and cloud providers. However, the way to request storage
resources or provision disks is different with the various cloud providers and
on-premises systems. This calls for a need to request storage resources in a
consistent manner across different hardware vendors and cloud providers.

Kubernetes solution to this is to split storage resources into two Kubernetes objects.
A PV is an object that defines a physical storage volume. It contains the details of
the underlying storage infrastructure. A PVC, on the other hand, is an abstracted
pointer to a PV. A PVC indicates that the owner has a claim on a specific PV. Pods
storage resources are associated with PVCs and never directly with the PV; this way,
the underlying storage definition is abstracted from the application.

The following diagram shows the relation between pods, PVCs, and PVs. The pod mounts a PVC as a volume; the PVC works as an abstraction layer for the pod to request a physical volume to be associated with the pod; the PVC is bound to a PV that provides specifics of the disks:

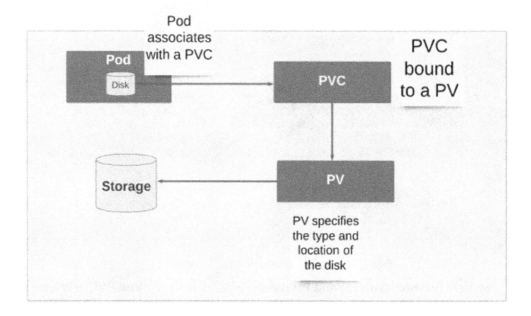

Figure 3.4 – Storage provisioning in Kubernetes (continued)

- **Ingress**: Services enable access to pods within the Kubernetes cluster. For scenarios in which you need access to a pod from outside the Kubernetes cluster, Ingress is the answer. Ingress provides a way for you to expose a particular service to be accessible from outside the cluster. This enables you to map a **HyperText Transfer Protocol** (**HTTP**)-based **Uniform Resource Locator** (**URL**) that points to a service. Ingress may also use **Secure Sockets Layer** (**SSL**) on the exposed URL and can be configured to terminate SSL for traffic within the cluster. This way, the transport layer will be encrypted all the way up to the Ingress, while forwarding the traffic to the pod in plain HTTP. It is also worth noting that Kubernetes allows traffic to be encrypted all the way to the pod if needed.

The following diagram shows how Ingress enables pods to be accessible from outside the Kubernetes cluster:

Figure 3.5 – The Ingress object in the Kubernetes cluster

Now that you have refreshed your understanding of Kubernetes, let's see how Kubernetes allows you to run your platform anywhere.

Becoming cloud-agnostic through Kubernetes

One of the key aspects of the ML platform we are building is that it enables the organization to run on any cloud or data center. However, each cloud has its own proprietary APIs to manage resources and deploy applications. For example, the **Amazon Web Services** (**AWS**) API uses an **Elastic Compute Cloud** (**EC2**) instance (a server) when provisioning a server, while Google Cloud's API uses a **Google Compute Engine** (**GCE**) VM (a server). Even the names of the resources are different! This is where Kubernetes plays a key role.

The wide adoption of Kubernetes has forced major cloud vendors to come up with tight integration solutions with Kubernetes. This allows anyone to spin up a Kubernetes cluster in AWS, GCP, or Azure in a matter of minutes.

The Kubernetes API enables you to manage cloud resources. Using the standard Kubernetes API, you can deploy applications on any major cloud provider without needing to learn about the cloud provider's API. The Kubernetes API has become the abstraction layer to manage workloads in the cloud. The ML platform you will build in this book will exclusively use Kubernetes APIs to deploy and run applications. This includes the software components that make up the ML platform.

The following diagram shows how Kubernetes allows you to become cloud-agnostic. You interact with Kubernetes through the Kubernetes API, which eliminates or reduces the need to interact directly with the cloud vendor's API. In other words, Kubernetes provides a consistent way of interacting with your environment regardless of which cloud or data center it is running on:

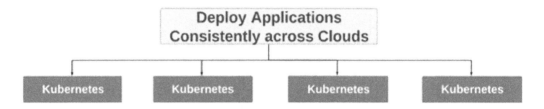

Figure 3.6 – Kubernetes acting as a shim to cloud providers APIs

Another important thing that came out of the Kubernetes community is **Operators**. You will be using Kubernetes Operators to deploy most of the components of the ML platform. Let's dig in.

Understanding Operators

In traditional **information technology** (**IT**) organizations, specialized and dedicated teams were required to maintain applications and other software components such as databases, caches, and messaging components. Those teams were continuously observing the software ecosystem and doing specific things such as taking backups for databases, upgrading and patching newer versions of software components, and more.

Operators are like system administrators or engineers, continuously monitoring applications running on the Kubernetes environment and performing operational tasks associated with the specific component. In other words, an Operator is an automated software manager that manages the installation and life cycle of applications on Kubernetes.

Put simply, instead of you creating and updating Kubernetes Objects (Deployment, Ingress, and so on), the Operator takes this responsibility based on the configuration you provide. The configuration that directs the Operator to perform certain tasks is called a **custom resource** (**CR**), and the structure or schema for a CR is defined by an object called a **CR definition** (**CRD**).

The following diagram shows how an Operator automates application operations activities. In the traditional approach, the developer builds and develops the application, and then an application operations team provides support to run the application. One of the Kubernetes Operator's aims is to automate activities that operations people perform:

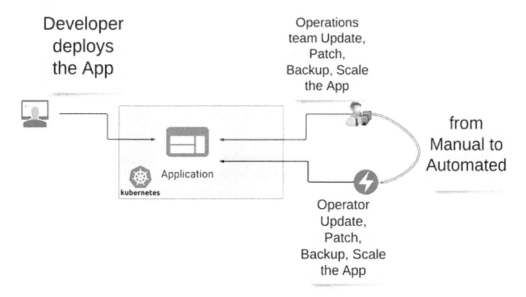

Figure 3.7 – An Operator is a software that automates tasks of the operations team

Kubernetes Operators can be complex. There are Operators that manage instances of databases, while some manage clusters of pods that work together. Some Operators own just 1 or 2 CRDs, while others could own more than 10 CRDs. The **Operator Lifecycle Manager** (**OLM**) simplifies the installation and management of Kubernetes Operators. Let's dig a little bit deeper into this.

In OLM, there are multiple stages required to install an Operator: creating a Deployment object for the Operator, configuring the required permissions to run an Operator (because it needs to observe changes in the Kubernetes cluster), and creating a CRD. To reduce the complexity of installing an Operator, a management layer may come in handy. OLM fulfills this role.

OLM standardizes interactions with Operators. It requires that all interactions with the Operator be done through the Kubernetes API. OLM makes it easy to manage the life cycle of multiple Operators through a single standard interface—the Kubernetes API. Our ML platform will make use of a few Operators, and therefore it is useful to understand OLM and objects related to it. Let's look at them in more detail here:

- `ClusterServiceVersion`: This object defines metadata about an Operator. It includes the name and version of the Operator, along with the installation information and required permissions. It also describes the CRD owned and required by the Operator.

- `Subscription`: The `Subscription` object allows the user to install and update the Operator. OLM uses this object to install and configure Operators, CRDs, and related access-control objects.

- `OperatorGroup`: `OperatorGroup` provides a way to associate your Operator with a specific set of namespaces. `OperatorGroup` defines a set of namespaces to which the associated Operator will react. If we do not define a set of namespaces in the `OperatorGroup` definition, then the Operator will run globally across all namespaces.

In the next section, you will get to install and configure your local Kubernetes environment and install OLM on the Kubernetes cluster.

Setting up your local Kubernetes environment

Now that we have refreshed some basic Kubernetes concepts, it's time for the rubber to hit the road. In this section, we will prepare and validate our local Kubernetes clusters. The cluster we set up here will be used to host the ML platform in later chapters.

Installing kubectl

`kubectl` is a command-line tool that assists in running commands against a Kubernetes cluster. You can create Kubernetes objects, view logs, and monitor the progress of your actions through this utility. The following steps will install `kubectl` on your machine.

Installing kubectl on Linux

First, let's see the process for installing `kubectl` on a machine running Linux. Follow these next steps:

1. Create or **Secure Shell** (**SSH**) to a Terminal session on your Linux computer.

2. Download the `kubectl`. Kubernetes **command-line interface** (**CLI**). We will be using version `1.22.4` throughout the book. The following two lines of code are one command:

    ```
    curl -LO https://dl.k8s.io/release/v1.22.4/bin/linux/
    amd64/kubectl
    ```

3. Install the `kubectl` CLI by running the following command:

    ```
    sudo install kubectl /usr/local/bin/kubectl
    ```

4. Validate that it is installed by running the following command:

    ```
    kubectl version --client
    ```

    You should see the following response to the `version` command:

```
$kubectl version —client
Client Version: version.Info{Major:"1", Minor:"22", GitVersion:"v1.22.4",
sion:"go1.16.10", Compiler:"gc", Platform:"linux/amd64"}
```

Figure 3.8 – Output of the kubectl version command in Linux

You should now have `kubectl` running on your Linux machine.

Installing kubectl on macOS

First, let's see the process for installing `kubectl` on a machine running macOS. Follow these steps:

1. Create or `SSH` to a Terminal session on your Mac computer.

2. Download the `kubectl` Kubernetes CLI. We will be using version *1.22.4* throughout the book.

    For Intel Macs, run the following command:

    ```
    curl -LO https://dl.k8s.io/release/v1.22.4/bin/darwin/
    amd64/kubectl
    ```

    For Apple M1 Macs, run the following command:

    ```
    curl -LO https://dl.k8s.io/release/v1.22.4/bin/darwin/
    aa64/kubectl
    ```

3. Install the `kubectl` CLI by running the following command:

    ```
    sudo install kubectl /usr/local/bin/kubectl
    ```

4. Validate that it is installed by running the following command:

```
kubectl version --client
```

You should see the following response to the `version` command:

```
$kubectl version –client
Client Version: version.Info{Major:"1", Minor:"22", GitVersion:"v1.22.4",
sion:"go1.16.10", Compiler:"gc", Platform:"linux/amd64"}
```

Figure 3.9 – Output of the kubectl version command in macOS

You should now have `kubectl` running on macOS.

Installing kubectl on Windows

Now, let's go through the steps for Windows, as follows:

1. Run **PowerShell** as **Administrator**.

2. Download the `kubectl` Kubernetes CLI binary by running the following command. We will be using version *1.22.4* throughout the book:

```
curl.exe -LO https://dl.k8s.io/release/v1.22.4/bin/
windows/amd64/kubectl.exe
```

3. Copy the `kubectl.exe` file to `c:\kubectl` by running the following commands:

```
mkdir c:\kubectl
copy kubectl.exe c:\kubectl
```

4. Add `c:\kubectl` to the PATH environment variable by running the following command and then restart your PowerShell Terminal:

```
setx $ENV:PATH "$ENV:PATH;C:\kubectl" /M
```

5. Validate that it is installed by running the following command:

```
kubectl version -client
```

You should see the following response to the `version` command:

```
PS C:\Windows\system32> kubectl version --client
Client Version: version.Info{Major:"1", Minor:"22", GitVersion:"v1.22.4", Git
e75f1", GitTreeState:"clean", BuildDate:"2021-11-17T15:48:33Z", GoVersion:"go
amd64"}
```

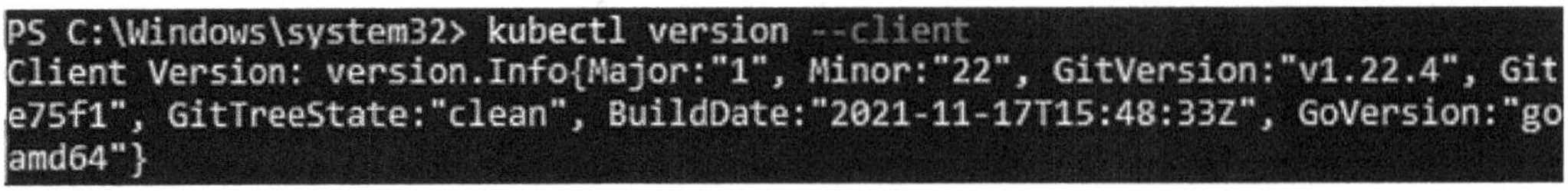

Figure 3.10 – Output of kubectl version command in Windows

You have just installed the kubectl Kubernetes CLI. The next step is to install minikube, a local, single-node Kubernetes cluster.

Installing minikube

minikube provides a way to run a local Kubernetes cluster with ease. This is a minimal cluster, and it is intended to be only used for local development or experimentation. Running Kubernetes in production environments is beyond the scope of this book.

As with kubectl, let's go through the installation for different types of operating systems.

Installing minikube on Linux

Follow these steps to install minikube on Linux:

1. Create a Terminal session or SSH to your Linux computer.

2. Install podman for minikube using the following code:

    ```
    sudo dnf install podman -y
    ```

3. Download minikube from this location. We are using version 1.24.0 of minikube:

    ```
    curl -LO https://storage.googleapis.com/minikube/
    releases/v1.24.0/minikube-linux-amd64
    ```

4. Install the minikube utility, as follows:

    ```
    sudo install minikube-linux-amd64 /usr/local/bin/minikube
    ```

5. Validate the minikube version, like this:

    ```
    minikube version
    ```

    You should see the following response:

    ```
    $minikube version
    minikube version: v1.24.0
    commit: 76b94fb3c4e8ac5062daf70d60cf03ddcc0a741b
    $
    ```

    Figure 3.11 – Output of the minikube version command on Linux

You have just installed kubectl and minikube on Linux. These two command-line tools will help you to set up a local Kubernetes cluster.

Installing minikube on macOS

Although our preferred operating system is Linux for this book, we are providing steps to install `minikube` on macOS too. A lot of developers use the macOS system, and it would be beneficial to provide details for the operating system from Apple. Follow these next steps:

1. Download and install Docker Desktop from the Docker website or by accessing the following web page: `https://www.docker.com/products/docker-desktop`.

2. Once Docker is available, make sure that it is installed correctly by running the following command. Make sure that Docker is running before running this command:

    ```
    docker version
    ```

    You should see the following response. If you get an error, please make sure that Docker is running:

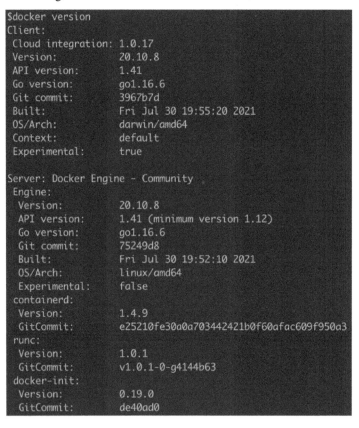

```
$docker version
Client:
 Cloud integration: 1.0.17
 Version:           20.10.8
 API version:       1.41
 Go version:        go1.16.6
 Git commit:        3967b7d
 Built:             Fri Jul 30 19:55:20 2021
 OS/Arch:           darwin/amd64
 Context:           default
 Experimental:      true

Server: Docker Engine - Community
 Engine:
  Version:          20.10.8
  API version:      1.41 (minimum version 1.12)
  Go version:       go1.16.6
  Git commit:       75249d8
  Built:            Fri Jul 30 19:52:10 2021
  OS/Arch:          linux/amd64
  Experimental:     false
 containerd:
  Version:          1.4.9
  GitCommit:        e25210fe30a0a703442421b0f60afac609f950a3
 runc:
  Version:          1.0.1
  GitCommit:        v1.0.1-0-g4144b63
 docker-init:
  Version:          0.19.0
  GitCommit:        de40ad0
```

Figure 3.12 – Output of the docker version command on macOS

3. Open a Terminal on your macOS computer.

4. Download `minikube` by running one of the following commands. You will be using version 1.24.0 of Minikube:

    - If you have an Intel Mac, run the following command:

        ```
        curl -Lo minikube https://storage.googleapis.com/
        minikube/releases/v1.24.0/minikube-darwin-amd64
        ```

    - If you have an M1 Mac (Apple silicon), run this command instead:

        ```
        curl -Lo minikube https://storage.googleapis.com/
        minikube/releases/v1.24.0/minikube-darwin-arm64
        ```

5. Move the downloaded file to the `/usr/local/bin` folder and make the downloaded file an executable by using the following commands:

    ```
    sudo mv minikube /usr/local/bin
    sudo chmod +x /usr/local/bin/minikube
    ```

6. Validate the `minikube` version, as follows:

    ```
    minikube version
    ```

    You should see the following response:

    ```
    $minikube version
    minikube version: v1.24.0
    commit: 76b94fb3c4e8ac5062daf70d60cf03ddcc0a741b
    ```

Figure 3.13 – Output of the minikube version command

You have just installed `kubectl` and `minikube` on macOS. These two command-line tools will help you set up a local Kubernetes cluster.

Installing minikube on Windows

As with macOS, a substantial number of developers use Windows. It would be fair to provide steps on how to run the exercises on the operating system from Microsoft, the mighty Windows. Let's dig in on how to run `minikube` on Windows using `Hyper-V`, the Microsoft virtualization layer. Please note that `Hyper-V` is available on all Windows except Windows Home. Follow these steps:

1. Run **PowerShell** as **Administrator**.

2. In the PowerShell console, run the following command to enable `Hyper-V`:

    ```
    Enable-WindowsOptionalFeature -Online -FeatureName
    Microsoft-Hyper-V --All
    ```

    You should see the following response if `Hyper-V` is not enabled. If it is enabled already, the command will just print the status. Press *Y* to continue:

    ```
    Do you want to restart the computer to complete this operation now?
    [Y] Yes  [N] No  [?] Help (default is "Y"): _
    ```

    Figure 3.14 – Output of the command for enabling Hyper-V on Windows

    Restart the computer, if needed.

3. Download the `minikube` installer by opening the following link in the browser: `https://github.com/kubernetes/minikube/releases/download/v1.24.0/minikube-installer.exe`.

4. Run the downloaded installer. You should see the language setup screen, as shown in the following screenshot. Click **OK**:

Figure 3.15 – Language selection dialog of the minikube installer

5. The installer will present the following welcome screen. Click **Next >**, as illustrated in the following screenshot:

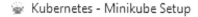 Kubernetes - Minikube Setup   —   ☐   ✕

Welcome to the Kubernetes - Minikube Setup Wizard

This wizard will guide you through the installation of Kubernetes - Minikube.

It is recommended that you close all other applications before starting Setup. This will make it possible to update relevant system files without having to reboot your computer.

Click Next to continue.

Next >   Cancel

Figure 3.16 – The minikube installer wizard

6. The installer will present the following **License Agreement** screen. Click **I Agree**:

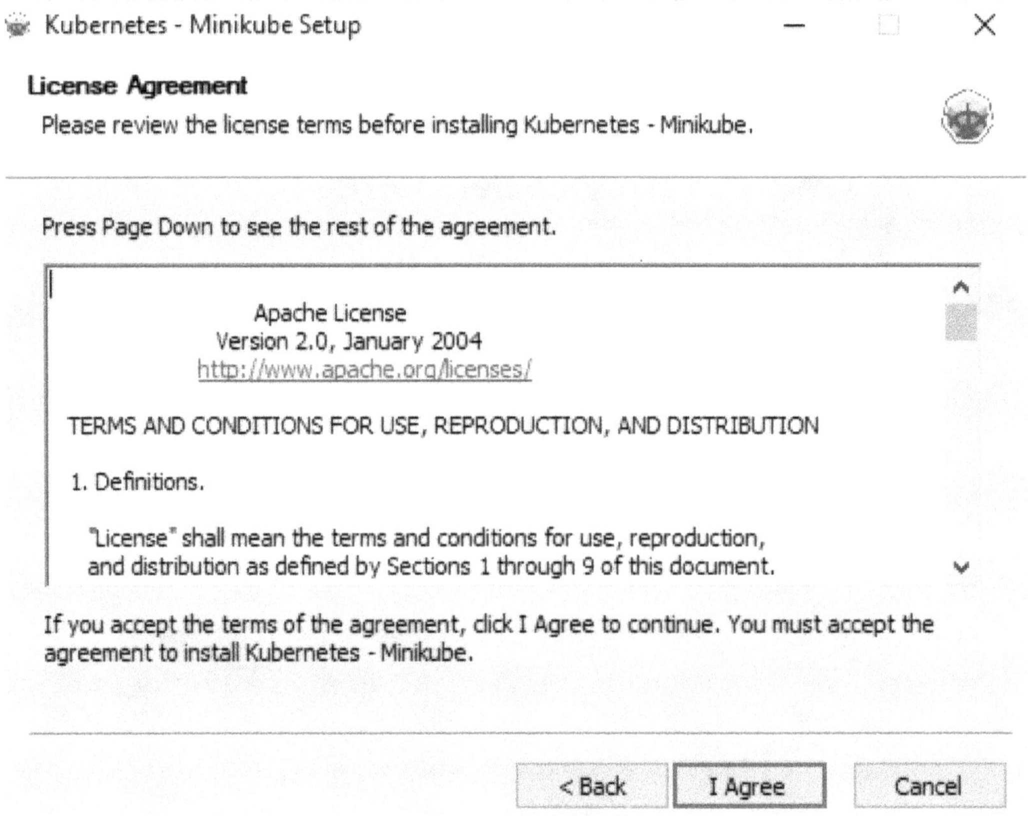

Kubernetes - Minikube Setup   —   □   ✕

License Agreement
Please review the license terms before installing Kubernetes - Minikube.

Press Page Down to see the rest of the agreement.

```
                        Apache License
                    Version 2.0, January 2004
                  http://www.apache.org/licenses/

  TERMS AND CONDITIONS FOR USE, REPRODUCTION, AND DISTRIBUTION

  1. Definitions.

    "License" shall mean the terms and conditions for use, reproduction,
    and distribution as defined by Sections 1 through 9 of this document.
```

If you accept the terms of the agreement, click I Agree to continue. You must accept the
agreement to install Kubernetes - Minikube.

< Back   I Agree   Cancel

Figure 3.17 – License Agreement screen of the minikube installer

7. On this screen, select the location where you want to install `minikube` and then click **Install**, as illustrated in the following screenshot:

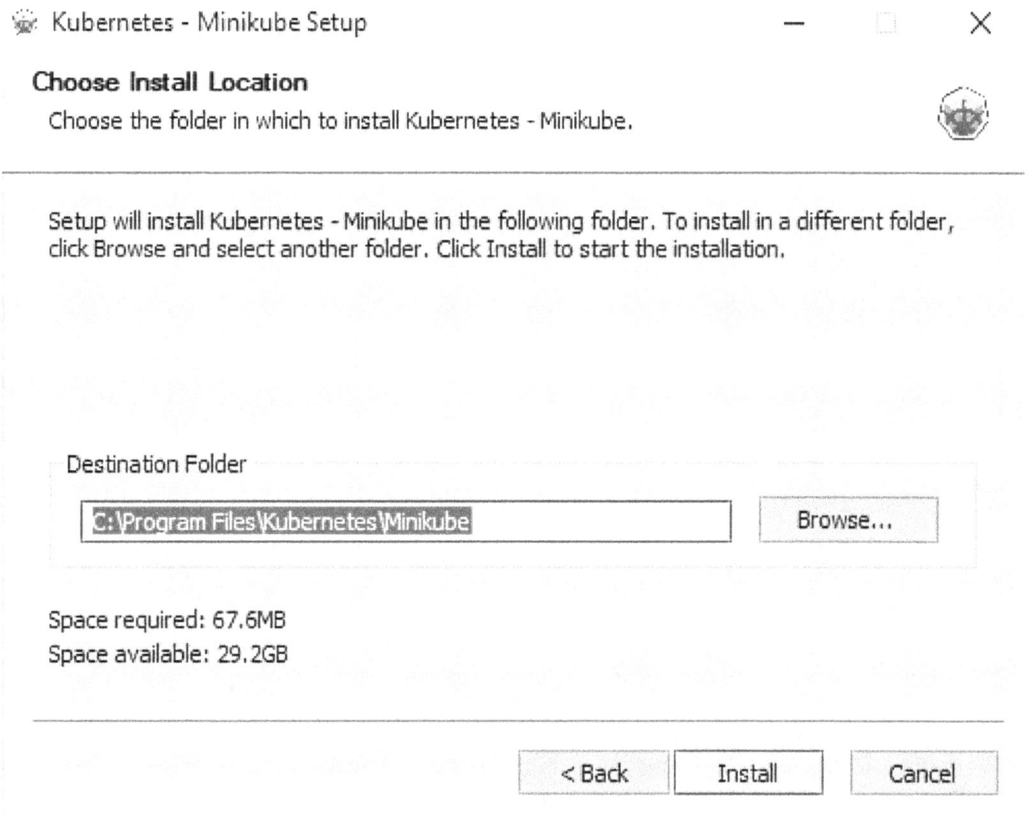

Figure 3.18 – Install location screen of the minikube installer

8. The installation may take a few minutes. Once the installation is successful, you should see the following screen. Click **Next >**:

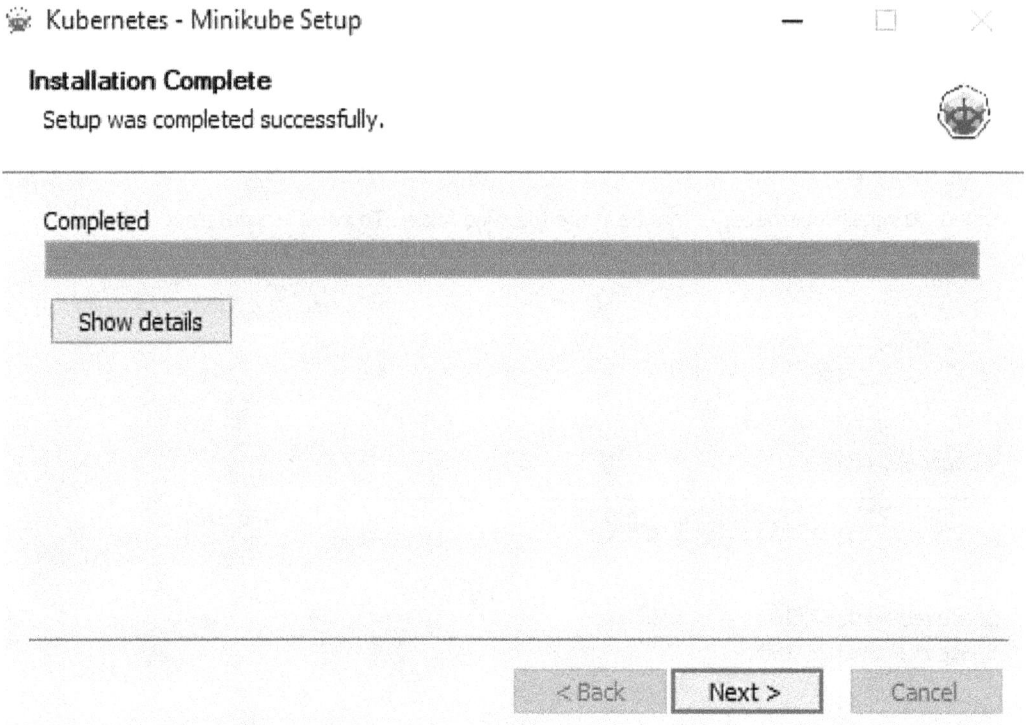

Figure 3.19 – Successful installation screen of the minikube installer

9. This is the last screen in your `minikube` setup process. Click **Finish** to complete it:

Figure 3.20 – Final screen of the minikube installer

10. Finally, in the PowerShell console, set the virtualization driver for `minikube` to `hyperv`. You can do this by running the following command:

```
minikube config set driver hyperv
```

You should see the following response:

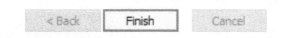

Figure 3.21 – Output of the minikube config command

Congratulations—you have set up the `minikube` program on your Windows machine!

Over the preceding sections, you have installed the `kubectl` and `minikube` tools to set up your Kubernetes cluster. In the next section, you will set up a Kubernetes cluster.

Setting up a local Kubernetes cluster

Now, we will set up a Kubernetes cluster on your local machine. As mentioned in the technical requirements, we will need a minimum of 4 CPU cores or **virtual CPUs (vCPUs)**, 60 GB of available disk, and at least 16 GB of memory to be allocated to the Kubernetes cluster. Our recommended configuration is 8 CPUs and 64 GB of memory with 60 GB of disk space. If you do not have these resources available locally, you can provision a Linux host in the cloud. We will describe in the next section how to provision a host on Google Cloud. Proceed as follows:

1. Set up the `minikube` configuration for CPU, disk, and memory through the following commands:

    ```
    minikube config set cpus 8
    minikube config set memory 32GB
    minikube config set disk-size 60GB
    ```

2. Validate if the configuration is set correctly via the following command:

    ```
    minikube config view
    ```

    You should see the following response:

    ```
    $minikube config view
    - disk-size: 60GB
    - memory: 64GB
    - cpus: 8
    ```

    Figure 3.22 – Output of the minikube config command

3. Now, start the Kubernetes cluster by running the following command:

    ```
    minikube start --kubernetes-version=1.22.4
    ```

    You should see the following response:

    ```
    $minikube start
    😊 minikube v1.24.0 on Fedora 35
    ```

    Figure 3.23 – Partial output of the minikube start command

    Once the start process is completed, you should see a successful message like this after the Kubernetes platform is available:

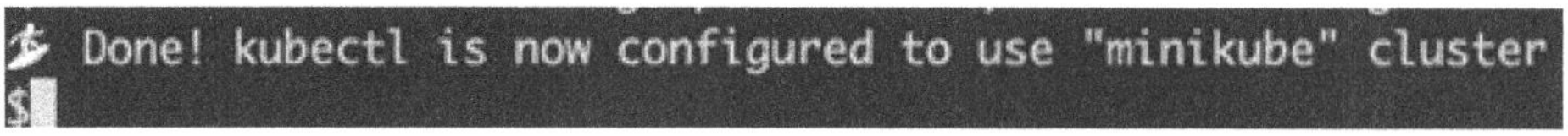

    ```
    ⚡ Done! kubectl is now configured to use "minikube" cluster
    $
    ```

    Figure 3.24 – Output after the successful start of minikube

4. Validate that all the pods are in the **Running** state through the following command on Linux or macOS. Note that it may take a few minutes for the pods to be in the *Running* state:

```
watch kubectl get pods --all-namespaces
```

Or, run this command in Windows PowerShell:

```
while (1) {kubectl get pods --all-namespaces; sleep 5}
```

You should see the following response:

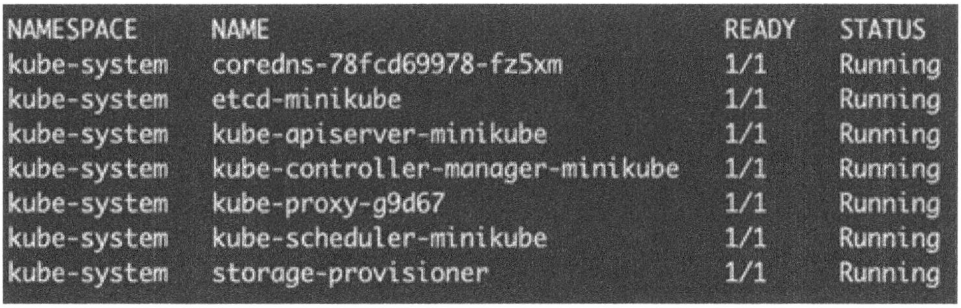

```
NAMESPACE     NAME                                      READY   STATUS
kube-system   coredns-78fcd69978-fz5xm                  1/1     Running
kube-system   etcd-minikube                             1/1     Running
kube-system   kube-apiserver-minikube                   1/1     Running
kube-system   kube-controller-manager-minikube          1/1     Running
kube-system   kube-proxy-g9d67                          1/1     Running
kube-system   kube-scheduler-minikube                   1/1     Running
kube-system   storage-provisioner                       1/1     Running
```

Figure 3.25 – Validating the Kubernetes pods have started successfully

Congratulations! You just installed and validated your new Kubernetes cluster. The next step is to install components that will allow you to run Operators on your new Kubernetes cluster.

Installing OLM

After you have validated that all pods are running for the local Kubernetes cluster, you will now install **OLM**. The process for installing OLM or any other applications inside Kubernetes is the same for all operating systems types. Proceed as follows:

1. Run the following command to install the CRD for the OLM:

```
kubectl apply -f https://github.com/operator-framework/
operator-lifecycle-manager/releases/download/v0.19.1/
crds.yaml
```

You should see the following response:

```
$kubectl apply -f https://github.com/operator-framework/operator-lifecycle-manager/releases/download/v0.19.1/crds.yaml

customresourcedefinition.apiextensions.k8s.io/catalogsources.operators.coreos.com created
customresourcedefinition.apiextensions.k8s.io/clusterserviceversions.operators.coreos.com created
customresourcedefinition.apiextensions.k8s.io/installplans.operators.coreos.com created
customresourcedefinition.apiextensions.k8s.io/operatorconditions.operators.coreos.com created
customresourcedefinition.apiextensions.k8s.io/operatorgroups.operators.coreos.com created
customresourcedefinition.apiextensions.k8s.io/operators.operators.coreos.com created
customresourcedefinition.apiextensions.k8s.io/subscriptions.operators.coreos.com created
$
```

Figure 3.26 – Validating OLM CRs have been created successfully

2. Run the following command to install OLM on Kubernetes:

```
kubectl apply -f https://github.com/operator-framework/
operator-lifecycle-manager/releases/download/v0.19.1/olm.
yaml
```

You should see the following response:

```
$kubectl apply -f https://github.com/operator-framework/operator-lifecycle-manager/releases/download/v0.19.1/olm.yaml
namespace/olm created
namespace/operators created
serviceaccount/olm-operator-serviceaccount created
clusterrole.rbac.authorization.k8s.io/system:controller:operator-lifecycle-manager created
clusterrolebinding.rbac.authorization.k8s.io/olm-operator-binding-olm created
deployment.apps/olm-operator created
deployment.apps/catalog-operator created
clusterrole.rbac.authorization.k8s.io/aggregate-olm-edit created
clusterrole.rbac.authorization.k8s.io/aggregate-olm-view created
operatorgroup.operators.coreos.com/global-operators created
operatorgroup.operators.coreos.com/olm-operators created
clusterserviceversion.operators.coreos.com/packageserver created
catalogsource.operators.coreos.com/operatorhubio-catalog created
```

Figure 3.27 – Creating OLM objects in Kubernetes

3. Validate if all OLM pods are in the *Running* state by running this command on Linux or macOS:

```
watch kubectl get pods -n olm
```

Or, run this command in Windows PowerShell:

```
while (1) {kubectl get pods -n olm; sleep 5}
```

You should see the following response:

```
NAME                                  READY   STATUS
catalog-operator-84976fd7df-7g98h     1/1     Running
olm-operator-844b4b88f8-ngqxj         1/1     Running
operatorhubio-catalog-5ztwv           1/1     Running
packageserver-6d87f5c89-8z6kv         1/1     Running
packageserver-6d87f5c89-qh5cr         1/1     Running
```

Figure 3.28 – Validating resources for OLM have been created successfully

4. Validate that `catalogsource` is available by issuing the following command:

```
kubectl get catalogsource -n olm
```

You should see the following response:

```
$kubectl get catalogsource -n olm
NAME                      DISPLAY               TYPE
operatorhubio-catalog     Community Operators   grpc
```

Figure 3.29 – Validating Operator catalog has been installed

Congratulations! You now have a local version of the Kubernetes cluster running and you have installed OLM on it. Your cluster is now ready to install Kubernetes Operators. Some of you may not have an access to a machine with the required minimum hardware requirements to run the ML platform, but don't worry—we've got you covered. The following section will help you provision the VM that you need in Google Cloud.

Provisioning a VM on GCP

It is always preferable to have a local environment that you can use to work on the exercises in this book. However, we understand that not everyone has the required compute resources available in their local machines. So, let's go to the cloud! You can provision just the right machine that you need for the exercises, in the cloud, and for free. For instance, Google Cloud gives **United States dollars** (**USD**) $300 worth of credit to new accounts. Other cloud providers such as AWS and Azure also give a similar free tier account, and it is up to you to select the cloud provider of your choice. For provisioning the VM we need for this book, however, we will use Google Cloud.

Once you have the account details sorted, use the following steps to provision a VM in your account. Just do not forget to stop the VM instance after you have completed a session to avoid getting billed for the hours that you are not using your machine.

The following instruction will guide you through the process of provisioning a VM in Google Cloud:

1. First, register for a new account at `https://cloud.google.com`.

2. Install the `gcloud` **software development kit** (**SDK**) by following the steps at `https://cloud.google.com/sdk/docs/install`.

3. Log in to Google Cloud using the following command. This command will open a browser instance where you provide login credentials for your Google Cloud account:

    ```
    gcloud auth login
    ```

    You should see the following response:

    ```
    $gcloud auth login
    Your browser has been opened to visit:
    ```

    Figure 3.30 – Output for the login command

4. Then, it will take you to the browser where you will authenticate. Once the browser completes the authentication steps, you will see the following output in the command line:

    ```
    You are now logged in as [masood.faisal@gmail.com].
    Your current project is [kube-test-258704].  You can change this setting by running:
     $ gcloud config set project PROJECT_ID
    ```

    Figure 3.31 – Output of a successful login to the gcloud account

5. Create a new project in Google Cloud, as follows. Your VM will belong to this project. Note that the project name must be globally unique in GCP, so please change it as per your preference:

    ```
    gcloud projects create mlops-kube --name="MLOps on
    Kubernetes"
    ```

    You should see the following response:

    ```
    $gcloud projects create mlops-kube --name="MLOps on Kubernetes"
    Create in progress for [https://cloudresourcemanager.googleapis.com/v1/projects/mlops-kube].
    Waiting for [operations/cp.6205518896565812931] to finish...done.
    Enabling service [cloudapis.googleapis.com] on project [mlops-kube]...
    Operation "operations/acf.p2-702800110954-740c89f8-e46f-44d6-a0d9-f42d0f254b2d" finished successfully.
    ```

    Figure 3.32 – Output of the create project command in Google Cloud

> **Projects in GCP**
>
> Project **identifiers** (**IDs**) or project names must be globally unique across Google Cloud. Only the first person will be able to create a project with the name mlops-kube. Choose a different project name of your choice for this command to work. You also need to use the chosen project name for subsequent commands where the mlops-kube project name is specified.

6. Make sure you are in the right project by issuing the following command:

```
gcloud config set project mlops-kube
```

You should see the following response:

```
$gcloud config set project mlops-kube
Updated property [core/project].
```

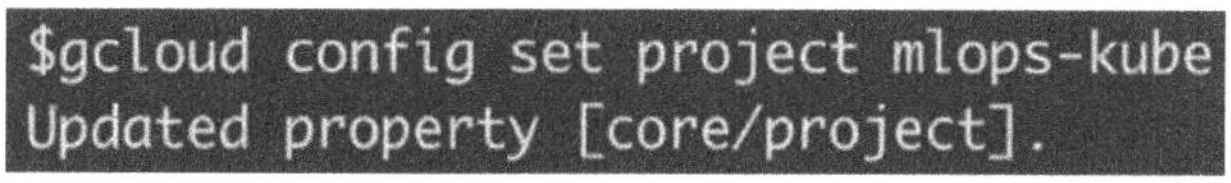

Figure 3.33 – Output of the command for setting the current project context

7. Set the right region and zone as per your location. You can get a list of all zones via the gcloud compute zones list command, as shown here:

```
gcloud config set compute/region australia-southeast1
```

You should see the following response:

```
$gcloud config set compute/region australia-southeast1
Updated property [compute/region].
```

Figure 3.34 – Output after setting up the gcloud region

Run the following command:

```
gcloud config set compute/zone australia-southeast1-a
```

You should then see the following response:

```
$gcloud config set compute/zone australia-southeast1-a
Updated property [compute/zone].
```

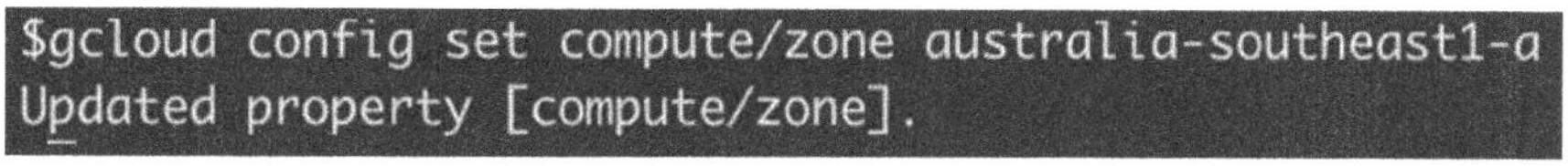

Figure 3.35 – Output after setting up the gcloud zone

8. Enable the Compute Engine API, as follows. This step is required to provision the Linux VM via APIs:

```
gcloud services enable compute.googleapis.com
```

9. Disable **OS Login** because you only connect via SSH, as follows:

```
gcloud compute project-info add-metadata --metadata
enable-oslogin=FALSE
```

10. Now, create a VM within this project by running the following command:

```
gcloud compute instances create mlopskube-cluster
--project=mlops-kube --zone=australia-southeast1-a
--machine-type=c2-standard-8 --network-interface=network-
tier=PREMIUM,subnet=default --maintenance-policy=MIGRATE
--service-account=702800110954-compute@developer.
gserviceaccount.com --scopes=https://www.googleapis.
com/auth/devstorage.read_only,https://www.googleapis.
com/auth/logging.write,https://www.googleapis.
com/auth/monitoring.write,https://www.googleapis.
com/auth/servicecontrol,https://www.googleapis.
com/auth/service.management.readonly,https://
www.googleapis.com/auth/trace.append --create-
disk=auto-delete=yes,boot=yes,device-name=instance-
1,image=projects/centos-cloud/global/images/centos-8-
v20211105,mode=rw,size=80,type=projects/mlops-kube/
zones/australia-southeast1-b/diskTypes/pd-balanced
--no-shielded-secure-boot --shielded-vtpm --shielded-
integrity-monitoring --reservation-affinity=any
```

The output of the command should display the machine details, as illustrated here:

NAME	ZONE	MACHINE_TYPE	PREEMPTIBLE	INTERNAL_IP	EXTERNAL_IP	STATUS
mlops-kubecluster	australia-southeast1-a	e2-highmem-8		10.152.0.2	34.151.100.35	RUNNING

Figure 3.36 – Output of the create VM command on Google Cloud

11. Add a firewall rule to allow access to the instance via port 22 for SSH, as follows. This is a lenient rule and should *NOT* be used in production:

```
gcloud compute --project=mlops-kube firewall-rules
create allow22 --direction=INGRESS --priority=1000
--network=default --action=ALLOW --rules=tcp:22 --source-
ranges=0.0.0.0/0
```

You should see the following response:

```
gcloud compute --project=mlops-kube firewall-rules create allow22 --direction=INGRESS --priority=1000 --network=default --action=ALLOW --rules=tcp:22 --source-ranges=0.0.0.0/0
Creating firewall...⠏Created [https://www.googleapis.com/compute/v1/projects/mlops-kube/global/firewalls/allow22].
Creating firewall...done.
NAME    NETWORK  DIRECTION  PRIORITY  ALLOW   DENY  DISABLED
allow22 default  INGRESS    1000      tcp:22        False
```

Figure 3.37 – Output of the firewall rule command

12. SSH to the machine using the `gcloud` SSH capability, as follows. This will give the command line, and you can call the Kubernetes command mentioned in the preceding section:

```
gcloud beta compute ssh --zone "australia-southeast1-a"
"mlopskube-cluster"  --project "mlops-kube"
```

13. Delete the instance after you have completed the session, as follows:

```
gcloud  compute instances delete --zone "australia-
southeast1-a" "mlopskube-cluster"  --project "mlops-kube"
```

You should see the following response:

```
$gcloud compute instances delete mlops-kubecluster
The following instances will be deleted. Any attached
flag is given and specifies them for keeping. Deleting
 - [mlops-kubecluster] in [australia-southeast1-a]

Do you want to continue (Y/n)?  Y
```

Figure 3.38 – Deleting the machine on Google Cloud

At this point, you can use this `gcloud` VM as the host machine for your Kubernetes cluster. Following the previous sections, you should now know how to install `kubectl` and `minikube` and how to set up a local Kubernetes cluster in this VM.

Summary

In this chapter, you have reviewed some basic Kubernetes concepts and gone through the Operator ecosystem in the Kubernetes universe. If you want to learn more about Kubernetes, *The Kubernetes Workshop* by Packt is a good place to start.

You have installed the tooling required to set up a local Kubernetes cluster. You have seen the instructions to do it in other environments such as Linux, macOS, and Windows. You have set up a VM on Google Cloud in case you do not want to use your local computer for the exercises. You have configured OLM to manage Operators on your Kubernetes cluster. These technologies will form the infrastructure foundation of our ML platform, which you will start to shape up in the next chapter.

Part 2: The Building Blocks of an MLOps Platform and How to Build One on Kubernetes

This part defines the different components of an MLOps solution in depth. The chapters provide details on each component and the purpose it serves. This section also provides an OSS solution that can play the role of each component in the MLOps platform.

This section comprises the following chapters:

- *Chapter 4, The Anatomy of a Machine Learning Platform*
- *Chapter 5, Data Engineering*
- *Chapter 6, Machine Learning Engineering*
- *Chapter 7, Model Deployment and Automation*

4

The Anatomy of a Machine Learning Platform

In this and the next few chapters, you will learn and install the components of a **machine learning** (**ML**) platform on top of Kubernetes. An ML platform should be capable of providing the tooling required to run the full life cycle of an ML project as described in *Chapter 2, Understanding MLOps*. This chapter starts with defining the different components of an ML platform in a technology-agnostic way. In the later parts, you will see the group of open source software that can satisfy the requirements of each component. We have chosen this approach to not tie you up with a specific technology stack; instead, you can replace components as you deem fit for your environment.

The solution that you will build in this book will be based on open source technologies and will be hosted on the Kubernetes platform that you built in *Chapter 3, Exploring Kubernetes.*

In this chapter, you will learn about the following topics:

- Defining a self-service platform
- Exploring the data engineering components
- Exploring the ML model life cycle components
- Addressing security, monitoring, and automation
- Exploring Open Data Hub

Technical requirements

This chapter includes some hands-on setup. You will be needing a running Kubernetes cluster configured with the **Operator Life cycle Manager** (**OLM**). Building such a Kubernetes environment is covered in *Chapter 3, Exploring Kubernetes*. Before attempting the technical exercises in this chapter, please make sure that you have a working Kubernetes cluster. You may choose to use a different flavor of Kubernetes than the one described in *Chapter 3, Exploring Kubernetes*, as long as the cluster has the OLM installed.

Defining a self-service platform

Self-service is defined as the capability of a platform that allows platform end users to provision resources on-demand without other human intervention. Take, for example, a data scientist user who needs an instance of a Jupyter notebook server, running on a host container with eight CPUs, to perform his/her work. A self-service ML platform should allow the data scientist to provision, through an end user friendly interface, the container that will run an instance of the Jupyter notebook server on it. Another example of self-service provisioning would be a data engineer requesting a new instance of an Apache Spark cluster to be provisioned to run his/her data pipelines. The last example is a data scientist who wants to package and deploy their ML model as a REST service so that the application can use the model.

One benefit of a self-service platform is that it allows cross-functional teams to work together with minimal dependencies on other teams. This independence results in better team dynamics, less friction, and increased team velocity.

The self-service model, however, needs governance. Imagine every data scientist requesting GPUs or data engineers requesting tens of terabytes of storage! Self-service capability is great, but without proper governance, it could also create problems. To avoid such problems, the platform has to be managed by a platform team that can control or limit the things the end users can do. One example of this limit is resource quotas. Teams and/or individual users can be allocated with quotas and be responsible for managing their own resources within the allocated quotas. Luckily, Kubernetes has this capability, and our ML platform can utilize this capability to apply limits to the team's resources.

As part of governance, the platform must have role-based access control. This is to ensure that only the users with the right role will have access to the resources they manage. For example, the platform team may be able to change the resource quotas, while data engineers can only spin up new Spark clusters and run data pipelines.

Another aspect of a self-service platform is the isolation of workloads. Many teams will be sharing the same platform and, while the quotas will keep the teams within their predefined boundaries, it is critical that there is a capability to isolate workloads from each other so that multiple unrelated projects running on the same platform do not overlap.

Exploring the data engineering components

In the context of this book, data engineering is the process of ingesting raw data from source systems and producing reliable data that could be used in scenarios such as analytics, business reporting, and ML. A data engineer is a person who builds software that collects and processes raw data to generate clean and meaningful datasets for data analysts and data scientists. These datasets will form the backbone for your organization's ML initiatives.

Figure 4.1 shows the various stages of a typical data engineering area of an ML project:

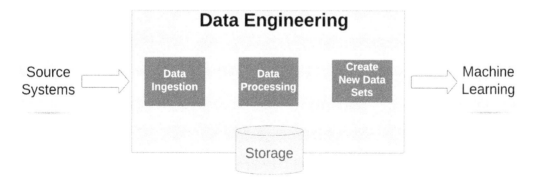

Figure 4.1 – Data engineering stages for ML

Data engineering often overlaps with **feature engineering**. While a data scientist decides on which features are more useful for the ML use case, he or she may work with the data engineer to retrieve particular data points that are not available in the current feature set. This is the main collaboration point between data engineers and data scientists. The datasets created by the data engineer in the data engineering block become the feature set in the ML block.

An ML platform that enables teams to perform feature engineering will have the following components and processes.

- **Data ingestion**: Data ingestion is the process in which the team understands the data sources and builds and deploys software that collects data from one or more data sources. Data engineers understand the impact of reading data from source systems. For example, while reading data from a source, the performance of the source system may get affected. Therefore, it is important for the ML platform to have a workflow scheduling capability so that the data collection can be scheduled during a time when the source system is less active.

  An ML platform enables the team to ingest data from various sources in multiple ways. For example, some data sources would allow data to be pulled, while other data sources may be able to push data. Data may come from a relational database, data warehouses, data lakes, data pools, data streams, API calls, or even from a raw filesystem. The platform should also have the capability to understand different protocols, for example, a messaging system may have multiple protocols, such as **Advanced Message Queuing Protocol** (**AMQP**), **Message Queuing Telemetry Transport** (**MQTT**), and Kafka. In other words, the ML platform should have the capability to gather data of various shapes and sizes from different types of data sources in various ways. *Figure 4.2* shows various sources of data from where the platform should be able to ingest the data:

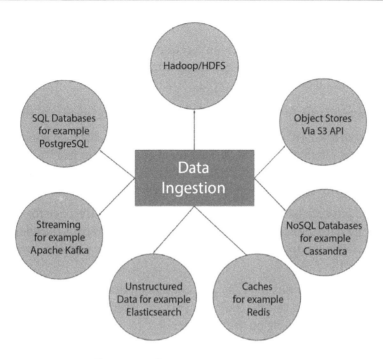

Figure 4.2 – Data ingestion integrations

- **Data transformation**: Once the data is ingested from various sources, it needs to be transformed from its original form into something that is more useful for the ML model training and other use cases. According to a Forbes survey, *80% of data scientists' work is related to preparing data for the model training*; this is the stage that is generally considered as boring among the data science teams. However, if the data is not transformed into an appropriate form, it will lead to less useful and/ or inefficient ML models. An ML platform enables teams to code, build, and deploy the data transformation pipelines and jobs with ease. The platform abstracts the complications of running and managing data transformation components such as Apache Spark jobs. Not only does the platform manage the execution of these processes, but it also manages the provisioning and cleaning of compute resources required to run these components, such as CPU, memory, and networking.

- **Storage**: During the feature engineering process, you will read and write data at various stages. You might create a temporary representation of the dataset for further processing, or you could write the new dataset to be used for ML processes. In these scenarios, you will need storage resources that can be accessed with ease and scale as needed. An ML platform provides on-demand storage for your datasets to be stored in a reliable fashion.

Now, let's see how the data engineer will use these components in their workflow.

Data engineer workflow

All the capabilities mentioned in the previous section are provided by the ML platform in a self-serving manner. The workflow that a data engineer using the platform would typically perform is as follows:

1. *Log in to the platform*: In this step, the data engineer authenticates to the platform.

2. *Provisioning of the development environment*: In this step, the data engineer requests the resource requirements for the development environment (such as the number of CPUs, amount of memory, and specific software libraries) to the platform. The platform then provisions the requested resources automatically.

3. *Build a data pipeline*: In this step, the data engineer writes the code for data ingestion and data transformation. The data engineer then runs the code in an isolated environment to verify its validity and perform the necessary refactoring and tuning of the code.

4. *Run a data pipeline*: In this step, the data engineer schedules the code to run as needed. It can be a regular schedule with periodic intervals such as hourly or daily, or a one-off run, depending on the use case.

You can see in the preceding steps that besides writing the code, all other steps are declarative. The data engineer's focus will be on building the code to ingest and transform data. All other aspects of the flow will be taken care of by the ML platform. This will result in improved efficiency and velocity for the team. The declarative capability of the platform will allow teams to standardize processes across your organization, which will reduce the number of bespoke toolchains and improve the security of the overall process.

The main output of the data engineering flow is a usable, transformed, and partially cleaned set of data that can be used to start building and training a model.

Exploring the model development components

Once the cleaned data is available, data scientists then go through the problem and try to determine what set of patterns would be helpful for the situation. The key here is that the data scientist's primary role is to find patterns in the data. Model development components of the ML platform explore data patterns, build and train ML models, and trial multiple configurations to find the best set of configurations and algorithms to achieve the desired performance of the model.

Within the course of model development, data scientists or ML engineers build multiple models based on multiple algorithms. These models are then trained using the data gathered and prepared from the data engineering flow. The data scientist then plays around with several hyperparameters to get different results from model testing. The result of such training and testing is then compared with each of the other models. These experimentation processes are then repeated multiple times until the desired results are achieved.

The experimentation phase will result in a selection of the most appropriate algorithm and configuration. The selected model will then be tagged for packaging and deployment.

Figure 4.3 shows the various stages of model development for an ML project:

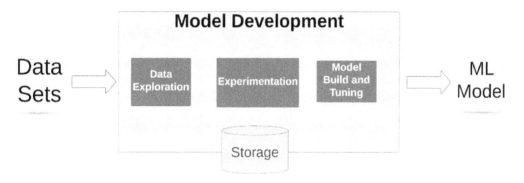

Figure 4.3 – Data engineering stages for ML

An ML platform that enables teams to perform model development will have the following components:

- **Data exploration**: We humans are better at finding patterns when the data is visualized as opposed to just looking at raw data sets. The ML platform enables you to visualize data. As a data scientist, you will need to collaborate with **subject matter experts** (**SMEs**) who have domain knowledge. Let's say you are analyzing a dataset of coronavirus patients. If you are not an expert in the virology or medicine domains, you will need to work with an SME who can provide insights about the dataset, the relationships of features, and the quality of the data itself. An ML platform allows you to share the visualizations you have created with the wider team for improved feedback. The platform also allows non-technical people to look at the data in a more graphical approach. This will help them gain a better understanding of the data.

- **Experimentation**: As a data scientist, you will split the data into training and testing sets, and then start building the model for the given metric. You will then experiment with multiple ML algorithms such as decision trees, XGBoost, and deep learning, and apply a variety of parameter tuning to each of the algorithms, for example, the number of layers or number of neurons for a deep learning model. This is what we call experimentation, and the platform enables the team to perform the experimentation in an autonomous way. Keep in mind that for each experiment, you may have different requirements for compute resources such as a GPU. This makes the self-service provisioning capability of the platform critical.

- **Tracking**: While doing multiple experiments, you need to keep track of the parameters used for each experiment and the metrics it has achieved. Some algorithms may require different sets of features, which means you also need to keep track of the version of the dataset that was used in training. There are two reasons for doing this. The first reason is that you will need a history of your experiments so that you can compare and pick the best combination. The second reason is that you may need to share the results with your fellow data scientists. The ML platform enables you to record the results of the experiments and share them seamlessly.

- **Model building and tuning**: In the experimentation stage, you have found the best algorithm and the best parameters for your model. You have compared the results and associated metrics for your model and have chosen the algorithm and parameters to be used. In this stage, you will train your model with these parameters, and register it with the model registry:

  - **Model registry**: As a data scientist, when you are satisfied with your model, you work with your team to deploy it. The real world changes, however, and you will need to update your model for new datasets or different metrics or simply for improved metrics. New versions of the models come all the time and the ML platform enables you to keep track of the versions of your models. The model versioning capability will help the team to compare the efficiency of new model versions with older model versions and allow the team to roll back a new model in production to previous versions if the need arises.

- **Storage**: Storage is not only important in the data engineering phase but also in model development. During the model development process, you read and write data at various stages. You split the dataset into a testing dataset and a training dataset, and you may choose to write it once so you can experiment with different model parameters but with the same datasets. The experiment tracking module and the model registry both need storage. The ML platform provides on-demand storage for your datasets to be stored in a reliable fashion.

Now, let's see how the data scientists use these components in their workflow.

Understanding the data scientist workflow

All the capabilities mentioned in the previous section are provided by the ML platform in a self-serving way. The typical workflow for the data scientist would be as follows:

1. *Log in to the platform*: The data scientists authenticate to the platform.

2. *Provisioning of the development environment*: In this step, the data scientist requests, to the platform, the resource requirements for the development environment, such as the number of CPUs, amount of memory, and specific software libraries. The platform then provisions the requested resources automatically.

3. *Exploratory data analysis*: In this stage, data scientists perform several types of data transformations and visualization techniques to understand the patterns hidden in the data.

4. *Experimenting with different algorithms*: In this stage, data scientists split the full dataset into training and testing sets. Then, the data scientists apply different ML algorithms and hyperparameters to achieve the desired metrics. Data scientists then compare the parameters of each training run to select the best one for the given use case.

5. *Model training*: Data scientists train the model as per the most optimized parameters found in the previous stage, and register the model in the model registry.

6. *Run model deployment pipeline*: In this step, the data scientists package the model to be consumed as a service and build the pipeline to automate the deployment process. It can be scheduled regularly or as a one-off run, depending on the use case.

You can see in the preceding steps that besides writing the code to facilitate model building and training, all other steps are declarative. The data scientists' focus will be on building more data science and ML engineering tasks. All other aspects of the flow will be taken care of by the ML platform. This will result in improved efficiency and velocity for the team, not to mention a happier data scientist. The declarative capability of the platform will also allow teams to standardize processes across your organization, which will reduce the use of bespoke toolchains improving consistency and improving the security of the overall process.

In the next section, you will explore the common services of the ML platform. These services are critical to making the platform production-ready and easier to adopt in the enterprise environment.

Security, monitoring, and automation

In this section, you will see some common components of the ML platform that apply to all the components and stages we have discussed so far. These components assist you in operationalizing the platform in your organization:

- **Data pipeline execution**: The outcome of data engineering is a data pipeline that ingests, cleans, and processes data. You have built this pipeline with scaled-down data for development purposes. Now, you need to run this code with production data, or you want a scheduled run with new data available, say, every week. An ML platform allows you to take your code and automate its execution in different environments. This is a big step because the platform not only allows you to run your code but will also manage the packaging of all the dependencies of your code so that it can run anywhere. If the code that you have built is using Apache Spark, the platform should allow you to automate the process of provisioning a Spark cluster and all other components required to run your data pipeline.

- **Model deployment**: Once the model is ready to be used, it should be available to be consumed as a service. Without the automated model packaging and deployment capability of the ML platform, the process of packaging a model and hosting it as a service requires some software engineering work. This work requires tight collaboration with software engineers and the operations team and may take days, if not weeks, to accomplish. The ML platform automates this process and it usually takes only a few seconds to a few minutes. The result of this process is an ML model deployed in an environment and is accessible as a service – typically, as a REST API.

  Deployment of the model is one aspect; over time, you may also need to re-train the model with new datasets. The platform also enables your team to automate the retraining process using the same training code you built for the first time when you trained your model. The retrained model is then redeployed automatically. This capability massively improves the efficiency of the team and this allows for more efficient use of time, such as working on newer challenges while delivering values for the business.

- **Monitoring**: Monitoring does not just refer to having the capability to observe the dynamics of the components in production, such as monitoring the model response time, but it also enables the team to respond to events before they become problems. A good monitoring platform provides observability during the full ML project life cycle and not just monitoring in production. When you are writing code to process data, you may need to tune the joins expression between datasets from multiple systems. This is one of the examples of information you need during development. The ML platform allows you to dig into the details during the development process. The platform also provides capabilities to monitor the underlying IT infrastructure. For example, when you are running your code during the model training stage, the platform provides the metrics on hardware resource utilization.

- **Security and governance**: The platform you are building allows teams to work autonomously. Teams can use the tools in the platform to perform the work anytime. However, the question of who can access what and who can use which tools proves to be a challenge for many organizations. For this, the platform must have an access control capability and provide access to only authorized users. The security component of the platform allows the users to be authenticated and authorized through standard protocols such as **OAuth2** or **OpenID Connect**. You will be using open source components to bring authentication components to the platform. The platform also uses the Kubernetes namespace feature to provide workload isolation across different teams that are sharing the same cluster. Kubernetes also provides the capability to assign limits of hardware resources to be used by individual teams. These capabilities will enable teams to share the platform across many different units within your organization while providing well-defined isolation boundaries and hardware resource quotas.

- **Source code management**: When you build data pipelines or train your model, you write code. The platform provides capabilities to integrate with source code management solutions. **Git** is the default source code management solution integrated platform.

Now, let's move on to cover **Open Data Hub** (**ODH**).

Introducing ODH

ODH is an open source project that provides most of the components required by our ML platform. It comes with a Kubernetes operator and a curated set of open source software components that make up most of the ML platform. In this book, we will mainly use the ODH operator. There are also other components that we will be using in the platform that don't originally come with ODH. One good thing about the ODH operator is the ability to swap default components for another as you see fit for your case.

To build the platform, you will use the following components. In the next few chapters, you will learn about the details of each of these components and how to use them. For now, you just need to understand their purpose at a very high-level:

- **ODH operator**: A Kubernetes operator that manages the life cycle of different components of the ML platform. It controls and manages the installation and maintenance of the software components used in your ML platform.

- **JupyterHub**: Manages instances of Jupyter Notebook servers and their related resources.

- **Jupyter notebooks**: An **integrated development environment** (IDE) is the main data engineering and data science workspace in the platform. Data scientists and engineers will use these workspaces to write and debug code for both data engineering and ML workflows.

- **Apache Spark**: A distributed, parallel data processing engine and framework for processing large datasets. It provides a wide array of data ingestion connectors to consume data from a variety of sources.

- **Apache Airflow**: A workflow engine that automates the execution and scheduling of data pipelines and model deployment. Airflow orchestrates different components of your data pipelines.

- **Seldon Core**: A library for packaging and deploying ML models as a REST service. It also has the capability of monitoring the deployed models. It provides support for popular ML frameworks, which gives it the capability to wrap and package ML models built with frameworks such as TensorFlow, scikit-learn, XGBoost, and PyTorch, as REST services.

- **Prometheus and Grafana**: These two components provide the monitoring capabilities for our platform. Prometheus provides the metrics database to record telemetry data provided by the components of the platform, and Grafana provides the **graphical user interface** (GUI) to visualize the captured metrics.

- **Minio**: An object storage provider that is compatible with Amazon S3 APIs. The Minio component is not part of the ODH toolchain, but we will extend and configure the ODH operator to manage the Minio component on the ML platform.

- **MLFlow**: A component for tracking different model experiments and also serves as the model registry of the platform. The MLFlow component is not part of the ODH toolchain, but we will extend the ODH operator to manage the MLFlow component on the ML platform

You will also install an open source identity provider component. The goal for this component is to provide a common single sign-on feature for all the platform components. We will use **Keycloak** as the identity management system, but this could be swapped with an OAuth2-based system that may already exist in your case. Keycloak is not part of the ODH, and we will show you how to install it as a separate component on your Kubernetes cluster.

Figure 4.4 shows the major open source software that serves as the main components of the ML platform. The ODH extensibility model allows you to add or choose which products to use for which components as per the requirements. You can replace any of the components with other open source products of choice. However, for the exercises in this book, we will use the product listed here:

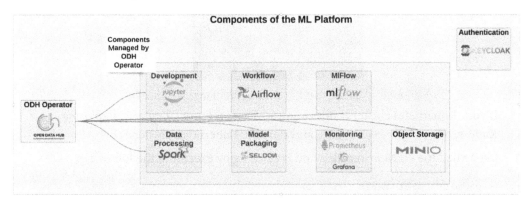

Figure 4.4 – Major components of the ML platform

In the next section, you will deploy the ODH operator and Keycloak server on your Kubernetes cluster. You will also install and configure the ingress controller to accept traffic from outside the cluster.

Installing the ODH operator on Kubernetes

In this section, you will install the ODH operator onto your Kubernetes cluster. At this stage, you will not enable any components of the platform. To install the operator, you first need to register the catalog source for the operator, and then you can install it.

First, let's register the catalog for the ODH operator. A catalog source contains metadata through which the OLM can discover operators and their dependencies. The ODH operator is not available in the default OLM catalog, so we need to register a new catalog that contains the ODH metadata for the OLM:

1. Validate that your Kubernetes cluster is running if you are using `minikube`:

    ```
    minikube status
    ```

    You should see the following response:

    ```
    $minikube status
    minikube
    type: Control Plane
    host: Running
    kubelet: Running
    apiserver: Running
    kubeconfig: Configured
    ```

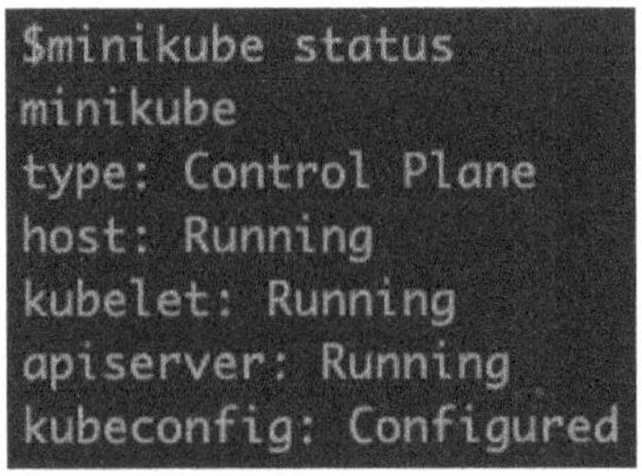

    Figure 4.5 – Validate that Kubernetes is running via minikube

    If your Kubernetes cluster is not running, please refer to *Chapter 3, Exploring Kubernetes*, on how to configure and start the Kubernetes cluster.

2. Verify that the OLM is installed and is running by executing the following:

    ```
    kubectl get pods -n olm
    ```

    You should see the following response:

    ```
    $kubectl get pods -n olm
    NAME                                READY   STATUS
    catalog-operator-84976fd7df-dc58n   1/1     Running
    olm-operator-844b4b88f8-ktfdx       1/1     Running
    operatorhubio-catalog-t7zb4         1/1     Running
    packageserver-69f6b89d6f-bw74p      1/1     Running
    packageserver-69f6b89d6f-f9lx2      1/1     Running
    ```

    Figure 4.6 – Command output showing OLM pods are running

    Make sure that all the OLM pods are running. If this is not the case for you, refer to *Chapter 3, Exploring Kubernetes*, in the *How to install OLM in your cluster* section.

3. Clone the Git repository and navigate to the repository's root directory. This repository contains all the source files, scripts, and manifests that you need to build the platform within the scope of this book: `https://github.com/PacktPublishing/Machine-Learning-on-Kubernetes.git cd Machine-Learning-on-Kubernetes`.

Register a new `catalog source` operator by using the YAML file available in the source code of this book:

```
kubectl create -f chapter4/catalog-source.yaml
```

4. After a couple of minutes, validate that the operator is available in your cluster:

```
kubectl get packagemanifests -o wide -n olm | grep -I
opendatahub
```

You should see the following response:

```
$kubectl get packagemanifests -o wide --namespace olm | grep -i OpenDataHub
opendatahub-operator                      Community Operators Red Hat    83s
$
```

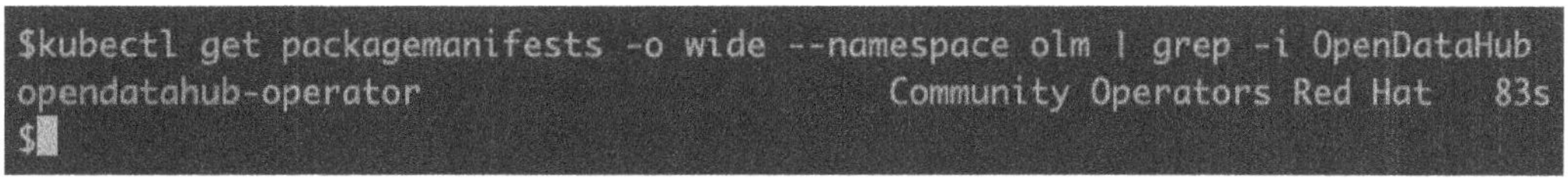

Figure 4.7 – Validate that the ODH operator is available

On Windows PowerShell, you may need to replace the `grep` command with `findstr`.

5. Now, create the subscription for the ODH operator. Recall from the third chapter that a subscription object triggers the installation of the operator via the OLM:

```
kubectl create -f chapter4/odh-subscription.yaml
```

You should see a response message that the subscription has been created.

6. After creating the subscription, the OLM will automatically install the operator and all its components. Verify that the ODH pod is running by issuing the following command. It may take a few seconds before the pods start appearing. If the pods are not listed, wait for a few seconds and rerun the same command:

```
kubectl get pods -n operators
```

You should see the following response:

```
$kubectl get pods -n operators
NAME                                      READY   STATUS    RESTARTS   AGE
opendatahub-operator-688bb984bf-s6vmg     1/1     Running   0          55s
```

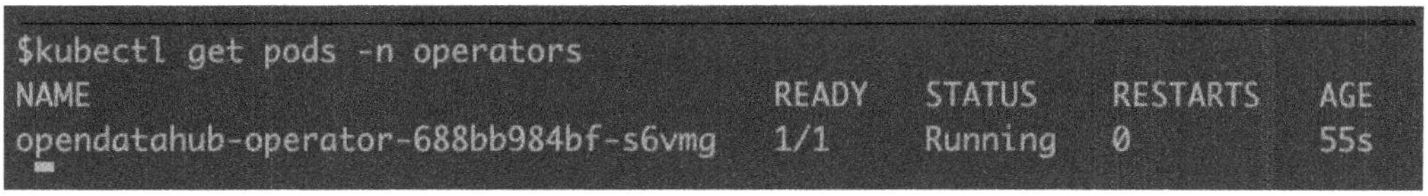

Figure 4.8 – Validate that the ODH pod is up and running

You have just installed the ODH operator on your Kubernetes cluster. Notice that you have not used generic Kubernetes objects such as **Deployments** to run your operator. The OLM allows you to easily manage the installation of an operator via the **Subscription** object.

In the next section, you install the ingress controller to allow traffic into your Kubernetes cluster.

Enabling the ingress controller on the Kubernetes cluster

Recall from *Chapter 3, Exploring Kubernetes*, that ingress provides a way for you to expose a particular service to make it accessible from outside the cluster. There are many ingress providers available on Kubernetes, and we leave it to you to select the right ingress provider for your cluster.

If you are using `minikube`, you need to follow these steps to enable the default ingress:

1. Enable the NGINX-based ingress controller for your cluster by issuing the following command:

```
minikube addons enable ingress
```

You should see the following response:

```
$minikube addons enable ingress --profile mlops
    ▪ Using image k8s.gcr.io/ingress-nginx/controller:v1.0.4
    ▪ Using image k8s.gcr.io/ingress-nginx/kube-webhook-certgen:v1.1.1
    ▪ Using image k8s.gcr.io/ingress-nginx/kube-webhook-certgen:v1.1.1
  ● Verifying ingress addon...
  ★ The 'ingress' addon is enabled
```

Figure 4.9 – Output for enabling minikube ingress plugin

2. Validate that the ingress pods are running in your cluster:

```
kubectl get pods -n ingress-nginx
```

You should see the following response:

```
$kubectl get pods -n ingress-nginx
NAME                                        READY   STATUS      RESTARTS   AGE
ingress-nginx-admission-create--1-xrntc     0/1     Completed   0          71s
ingress-nginx-admission-patch--1-bpss7      0/1     Completed   1          71s
ingress-nginx-controller-5f66978484-4rm57   1/1     Running     0          71s
$
```

Figure 4.10 – Validate that the Nginx ingress pods are in running state

Now that you have enabled the external traffic onto your cluster, the next step is to install the open source authentication and authorization component for your ML platform.

Installing Keycloak on Kubernetes

We will use Keycloak (`https://www.keycloak.org`) as our identity provider and add authentication and access management capabilities for your platform. Keycloak supports industry-standard security mechanisms such as **OAuth2** and **OpenID Connect**. In this section, you will install the Keycloak server on the Kubernetes cluster and log in to the Keycloak UI to validate the installation:

1. Start by creating a new namespace for the `keycloak` application:

    ```
    kubectl create ns keycloak
    ```

    You should see the following response:

    ```
    $kubectl create ns keycloak
    namespace/keycloak created
    ```

    Figure 4.11 – Output for creating a new namespace for Keycloak

2. Create the Keycloak manifest using the provided YAML file:

    ```
    kubectl create -f chapter4/keycloak.yaml --namespace
    keycloak
    ```

3. Validate that the `keycloak` pods are running. Note that the `--namespace` and `-n` flags are interchangeable in `kubectl`:

    ```
    kubectl get pods -n keycloak
    ```

    It may take a while to start, as it will start by pulling container images from the internet. The first time you run the command, you might see that the **READY** status is **0/1**. This is normal. Once the `Keycloak` pod is running, you should see the following response:

    ```
    $kubectl get pods -n keycloak
    NAME                          READY   STATUS    RESTARTS   AGE
    keycloak-6655d58b6c-f7rfg     1/1     Running   0          83s
    $
    ```

    Figure 4.12 – Validate that the Keycloak pods are in running state

    In the next few steps, you will define and configure the ingress for your Keycloak pod so that it can be accessed from outside the cluster.

4. Get the IP address of your `minikube` machine by issuing the following command:

```
minikube ip
```

You should see the following response:

```
$minikube ip
192.168.61.72
```

Figure 4.13 – IP address of your minikube instance

5. Open the `chapter4/keycloak-ingress.yaml` file and replace the `KEYCLOAK_` `HOST` string with the `keycloak.<THE_IP_ADDRESS_OF_YOUR_MINIKUBE>.` `nip.io` string. So, if the IP address of your `minikube` is `192.168.61.72`, then the string value would be `keycloak.192.168.61.72.nip.io`.

There are two places in the file where you need to put this new string. The file will look like *Figure 4.14*. Do not forget to save the changes in this file.

```yaml
apiVersion: networking.k8s.io/v1
kind: Ingress
metadata:
  name: keycloak
  namespace: keycloak
spec:
  tls:
    - hosts:
      - keycloak.192.168.61.72.nip.io
  rules:
    - host: keycloak.192.168.61.72.nip.io
      http:
        paths:
          - backend:
              service:
                name: keycloak
                port:
                  number: 8080
            path: /
            pathType: ImplementationSpecific
```

Figure 4.14 – The IP address of your minikube instance changed in the keycloak-ingress file

Apply the modified file to the Kubernetes cluster. This `ingress` object will create the required configuration for you to access the Keycloak server from outside the Kubernetes cluster. Run the following command to create the ingress object:

```
kubectl create -f chapter4/keycloak-ingress.yaml
--namespace keycloak
```

You should see the following response:

```
$kubectl create -f keycloak-ingress.yaml --namespace keycloak
ingress.networking.k8s.io/keycloak created
```

Figure 4.15 – Modified ingress has been applied

6. Validate that the `ingress` object is available by issuing the following command:

```
kubectl get ingress --namespace keycloak
```

You should see the following response:

```
$kubectl get ingress --namespace keycloak
NAME       CLASS   HOSTS                        ADDRESS     PORTS     AGE
keycloak   nginx   keycloak.192.168.61.72.nip.io  localhost   80, 443   64s
$
```

Figure 4.16 – Ingress object has been created

7. Now that you have validated that Keycloak is running and is exposed through the `ingress` object, open a browser on your machine where `minikube` is running and access the following URL. You need to replace the correct IP address, as stated in *step 5*: `https://keycloak.192.168.61.72.nip.io/auth/`.

You will get a warning that the *certificate is not valid*. This is because the Keycloak server uses a self-signed certificate by default. You just need to click the **Advance** button presented by the browser and choose to continue to the website.

You should see the following page; click on the **Administration Console** link to proceed further:

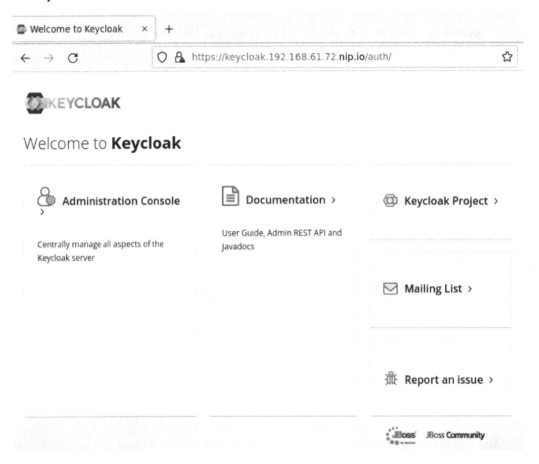

Figure 4.17 – Keycloak landing page

8. Log in using the credentials *admin/admin* in the following screen. After you enter the credentials, click **Sign in**:

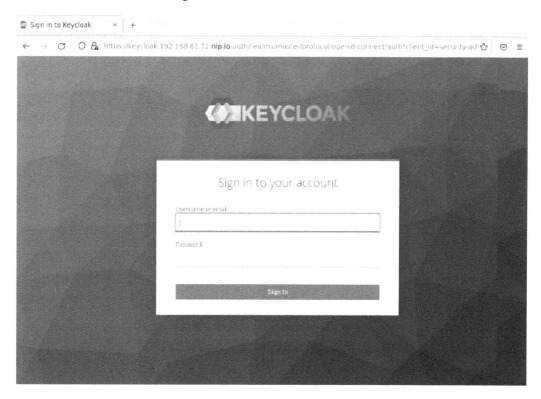

Figure 4.18 – Keycloak login page

9. Validate that the main administration page of Keycloak is displayed as follows:

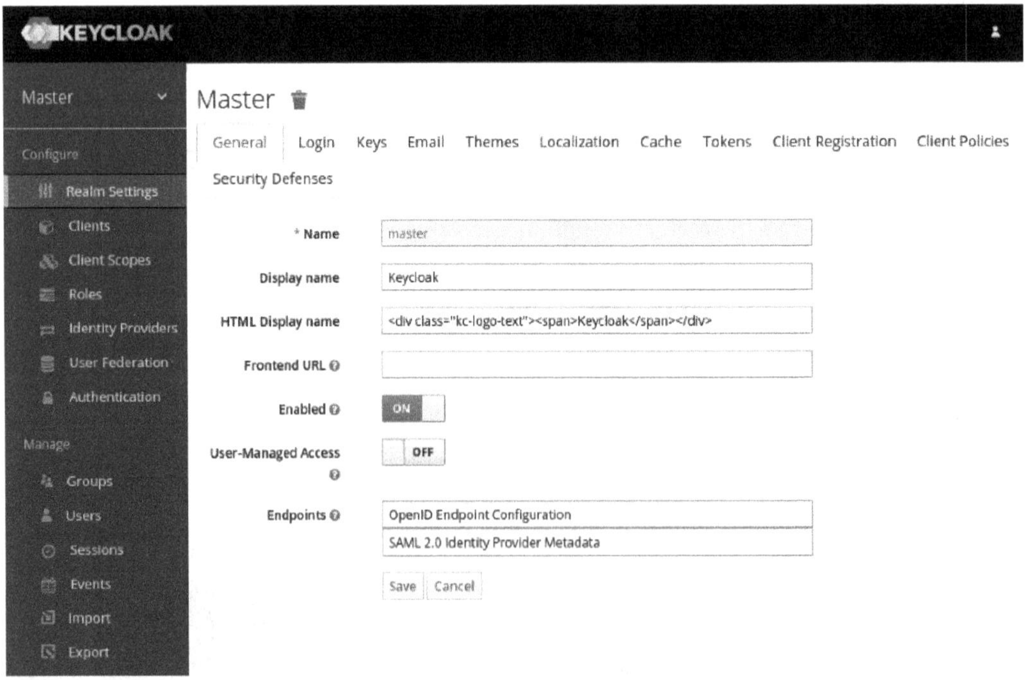

Figure 4.19 – Keycloak administration page

Congratulations! You have successfully installed the ODH operator and Keycloak onto your Kubernetes cluster.

Summary

In this chapter, you have learned about the major components of your ML platform and how open source community projects provide software products for each of those components. Using open source software enables a great number of people to use software for free, while at the same time, contributing to improving the components while continuously evolving and adding new capabilities to the software.

You have installed the operator required to set up the ML platform on your Kubernetes cluster. You have installed the ingress controller to allow traffic into your cluster and installed Keycloak to provide the identity and access management capabilities for your platform.

The foundation has been set for us to go deeper into each component of the ML life cycle. In the next chapter, you will learn to set up Spark and JupyterHub on your platform, which enables data engineers to build and deploy data pipelines.

Further reading

- Data preparation is the least enjoyable task in data science: `https://www.` `forbes.com/sites/gilpress/2016/03/23/data-preparation-` `most-time-consuming-least-enjoyable-data-science-task-` `survey-says/?sh=1e5986216f63`

5
Data Engineering

Data engineering, in general, refers to the management and organization of data and data flows across an organization. It involves data gathering, processing, versioning, data governance, and analytics. It is a huge topic that revolves around the development and maintenance of data processing platforms, data lakes, data marts, data warehouses, and data streams. It is an important practice that contributes to the success of **big data** and **machine learning** (**ML**) projects. In this chapter, you will learn about the ML-specific topics of data engineering.

A sizable number of ML tutorials/books start with a clean dataset and a CSV file to build your model against. The real world is different. Data comes in many shapes and sizes, and it is important that you have a well-defined strategy to harvest, process, and prepare data at scale. This chapter will discuss open source tools that can provide the foundations for data engineering in ML projects. You will learn how to install the open source toolsets on the Kubernetes platform and how these tools will enable you and your team to be more efficient and agile.

In this chapter, you will learn about the following topics:

- Configuring Keycloak for authentication

- Configuring Open Data Hub components

- Understanding and using the JupyterHub IDE

- Understanding the basics of Apache Spark

- Understanding how Open Data Hub provisions on-demand Apache Spark clusters

- Writing and running a Spark application from Jupyter Notebook

Technical requirements

This chapter includes some hands-on setup and exercises. You will need a running Kubernetes cluster configured with **Operator Lifecycle Manager**. Building such a Kubernetes environment is covered in *Chapter 3, Exploring Kubernetes*. Before attempting the technical exercises in this chapter, please make sure that you have a working Kubernetes cluster and **Open Data Hub** (**ODH**) installed on your Kubernetes cluster. Installing the ODH is covered in *Chapter 4, The Anatomy of a Machine Learning Platform*. You can find all the code associated with this book at `https://github.com/PacktPublishing/Machine-Learning-on-Kubernetes`.

Configuring Keycloak for authentication

Before you start using any component of your platform, you need to configure the authentication system to be associated with the platform components. As mentioned in *Chapter 4, The Anatomy of a Machine Learning Platform*, you will use Keycloak, an open source software to provide authentication services.

As a first step, import the configuration from `chapter5/realm-export.json`, which is available in the code repository associated with this book. This file contains the configuration required to associate the OAuth2 capabilities for the platform components.

Though this book is not a Keycloak guide by any means, we will provide some basic definitions for you to understand the high-level taxonomy of the Keycloak server:

- **Realm**: A Keycloak realm is an object that manages the users, roles, groups, and client applications that belong to the same domain. One Keycloak server can have multiple realms, so you have multiple sets of configurations, such as one realm for internal applications and one for external applications.

- **Clients**: Clients are entities that can request user authentication. A Keycloak client object is associated with a realm. All the applications in our platform that require **single sign-on** (**SSO**) will be registered as **clients** in the Keycloak server.

- **Users and groups**: These two terms are self-explanatory, and you will be creating a new user in the following steps and using it to log into different software of the platform.

The next step is to configure Keycloak to provide OAuth capabilities to our ML platform component.

Importing the Keycloak configuration for the ODH components

In this section, you will import the clients and group configurations onto the Keycloak server running on your Kubernetes cluster. The following steps will import everything onto the master realm of the Keycloak server:

1. Log in to your Keycloak server using the username `admin` and the password `admin`. Click on the **Import** link on the left-hand sidebar under the **Manage** heading:

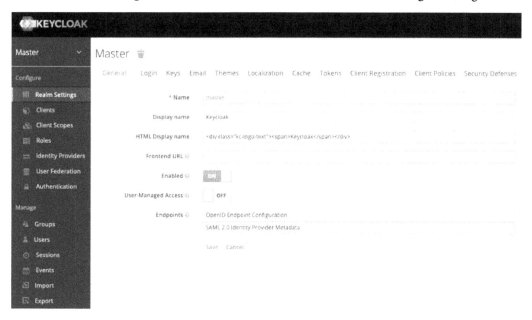

Figure 5.1 – Keycloak Master realm

2. Click on the **Select file** button on the screen, as follows:

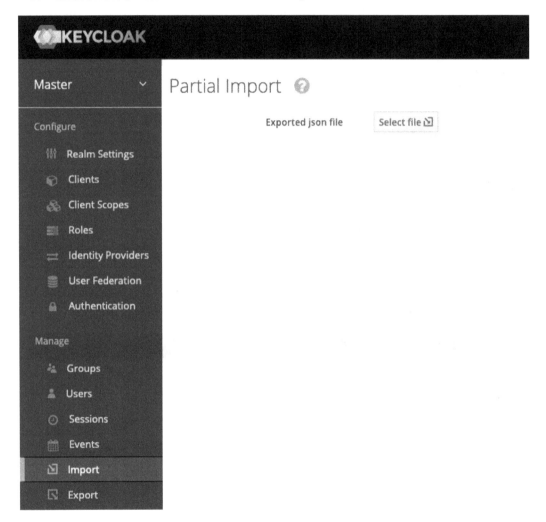

Figure 5.2 – Keycloak import configuration page

3. Select the `chapter5/realm-export.json` file from the pop-up window. After that, select **Skip** for the **If a resource exists** drop-down options, and click **Import**:

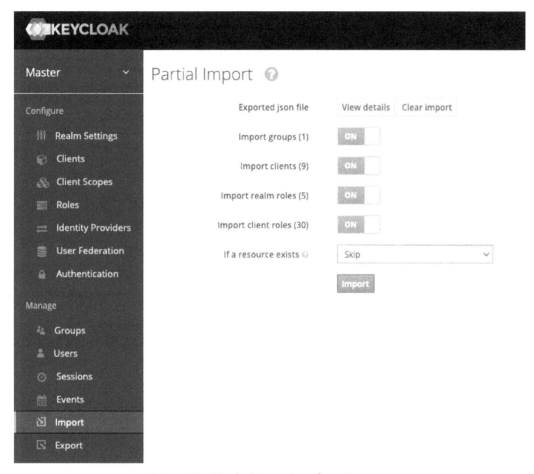

Figure 5.3 – Keycloak import configuration page

4. Validate that the records have been imported successfully onto your Keycloak server:

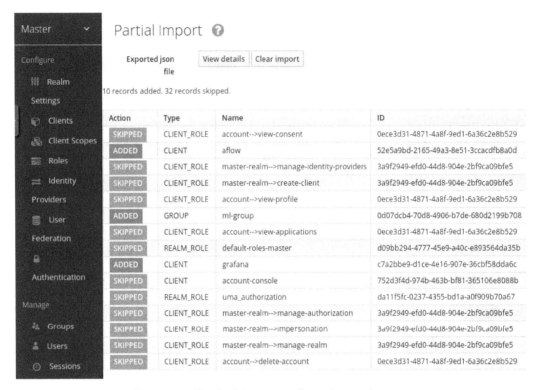

Figure 5.4 – Keycloak import configuration results page

5.   Validate that there are four clients created by clicking on the **Clients** item on the left-hand side menu. The following client IDs should exist: **aflow**, **mflow**, **grafana**, and **jhub**. The **aflow** client is for the workflow engine of the platform, which is an instance of **Apache Airflow**. The **mflow** client is for the model registry and training tracker tool and is an instance of **MLflow**. The **grafana** client is for monitoring UI and is an instance of **Grafana**. And last, the **jhub** client is for the **JupyterHub** server instance.

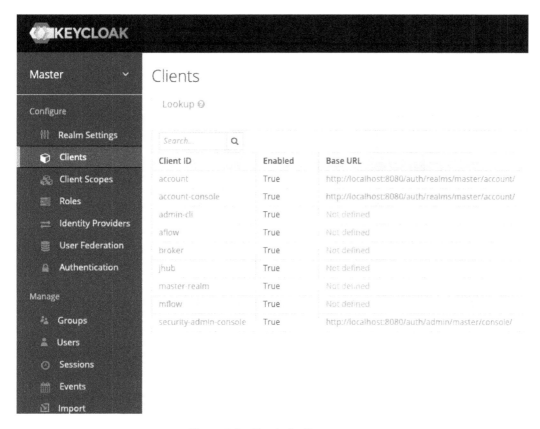

Figure 5.5 – Keycloak clients page

6. Validate that a group called **ml-group** has been created by clicking on the **Groups** link on the left-hand panel:

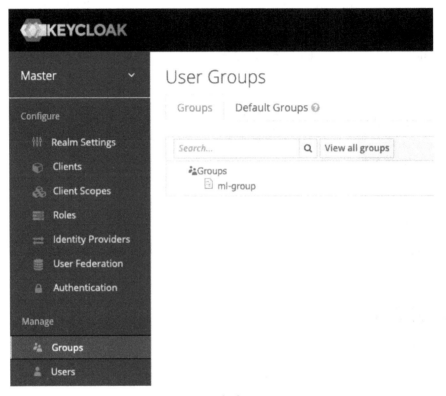

Figure 5.6 – Keycloak groups page

You will use this user group to create a user of the platform.

Great work! You have just configured multiple Keycloak clients for the ML platform. The next step is to create a user in Keycloak that you will be using for the rest of this book. It is important to note that Keycloak can be hooked with your enterprise directory or any other database and to use them as a source of the users. Keep in mind that the realm configuration we are using here is very basic and is not recommended for production use.

Creating a Keycloak user

In this section, you will create a new user and associate the newly created user with the group imported in the preceding section. Associating the user with the group gives the roles required for the different ODH software:

1. On the left-hand side of the Keycloak page, click on the **Users** link to come to this page. To add a new user, click the **Add user** button on the right:

Figure 5.7 – Keycloak users list

2. Add the username `mluser` and make sure the **User Enabled** and **Email Verified** toggle buttons are set to **ON**. In **Groups**, select the **ml-group** group and fill in the **Email**, **First Name**, and **Last Name** fields, as shown in *Figure 5.8*, and then hit the **Save** button:

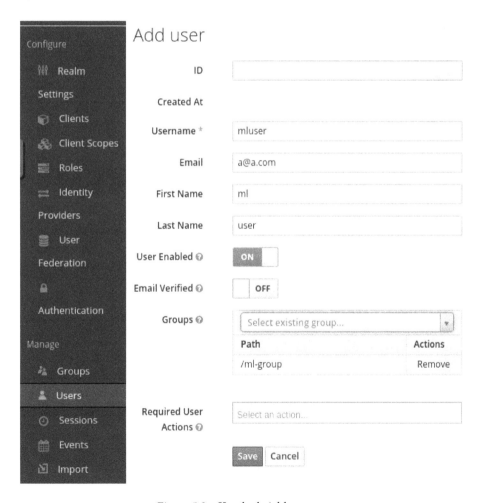

Figure 5.8 – Keycloak Add user page

3.   Click on the **Credentials** tab to set the password for your user:

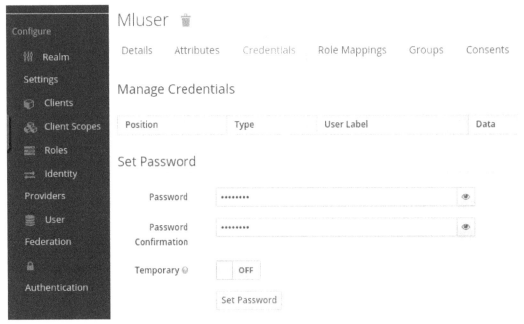

Figure 5.9 – Keycloak Credentials page

4.   Type in the password of your choice, then disable the **Temporary** flag, and hit the **Set Password** button.

You have just created and configured a user in Keycloak. Your Keycloak server is now ready to be used by the ML platform components. The next step is to explore the component of the platform that provides the main coding environment for all personas in the ML project.

Configuring ODH components

In *Chapter 4, The Anatomy of a Machine Learning Platform*, you have installed the ODH operator. Using the ODH operator, you will now configure an instance of ODH that will automatically install the components of the ML platform. ODH executes **Kustomize** scripts to install the components of the ML platform. As part of the code for this book, we have provided templates to install and configure all the components required to run the platform.

You can also configure what components ODH operators install for you through a `manifests` file. You can pass the specific configuration to the manifests and choose the components you need. One such manifest is available in the code repository of the book at `manifests/kfdef/ml-platform.yaml`. This YAML file is configured for the ODH operator to do its magic and install the software we need to be part of the platform. You will need to make some modifications to this file, as you will see in the following section.

This file defines the components of your platform and the location from where these components will get their settings:

- **Name**: Defines the name of the component.

- **repoRef**: This section contains the `path` property where you define the relative path location of the files required to configure this component.

- **Parameters**: This section contains the parameters that will be used to configure the component. Note that, in the following example, the IP address for `KEYCLOAK_URL` and `JUPYTERHUB_HOST` will need to be changed as per your configuration.

- **Overlays**: The ODH operator contains a default set of configurations for each component. Overlays provide a way to further tune the default configuration. The list of overlays is a set of folders, under the same location as the manifest file. The ODH operator will read the files from these overlay folders and merge them on the fly to produce a final configuration. You can find the overlays for JupyterHub in the `manifests/jupyterhub/overlays` folder in the code repository.

- **Repos**: This configuration section is specific to each manifest file and applies to all the components in the manifest. It defines the location and version of the Git repository that contains all the files being referred to by this manifest file. If you want the manifest to reference your own files for the installation, you need to refer here to the right Git repository (the repository that contains your files).

Figure 5.10 shows the part of the manifest file that holds the definition of the JupyterHub component:

```
- kustomizeConfig:
    overlays:
      - mlops
      - spark3

    parameters:
      - name: s3_endpoint_url
        value: "minio-ml-workshop.ml-workshop:9000"
      - name: KEYCLOAK_URL
        value: keycloak.192.168.61.72.nip.io
      - name: CLIENT_SECRET
        value: '30fc8258-459b-47d8-b81e-16aabe31ce4f'
      - name: JUPYTERHUB_HOST
        value: jupyterhub.192.168.61.72.nip.io

    repoRef:
      name: manifests
      path: manifests/jupyterhub/jupyterhub
  name: jupyterhub
```

Figure 5.10 – A component in the ODH manifest file

You will use the provided manifest file to create an instance of the ML platform. You may also tweak configurations or add or remove components of the platform as you wish by modifying this file. However, for the exercises in the book, we do not recommend changing this unless you are instructed to do so.

Now that you have seen the ODH manifest file, it's time to make good use of it to create your first ML platform on Kubernetes.

Installing ODH

Before we can install the data engineering components of the platform, we first need to create an instance of ODH. An ODH instance is a curated collection of related toolsets that serve as the components of an ML platform. Although the ML platform may contain components other than what is provided by ODH, it is fair to say that an instance of ODH is an instance of the ML platform. You may also run multiple instances of ODH on the same Kubernetes cluster as long as they run on their own isolated Kubernetes namespaces. This is useful when multiple teams or departments in your organization are sharing a single Kubernetes cluster.

The following are the steps you need to follow to create an instance of ODH on your Kubernetes cluster:

1. Create a new namespace in your Kubernetes cluster using the following command:

   `kubectl create ns ml-workshop`

   You should see the following response:

   ```
   $kubectl create ns ml-workshop
   namespace/ml-workshop created
   $
   ```

   Figure 5.11 – New namespace in your Kubernetes cluster

2. Make sure that the ODH operator is running by issuing the following command:

   `kubectl get pods -n operators`

   You should see the following response. Make sure the status says Running:

   ```
   $kubectl get pods -n operators
   NAME                                      READY   STATUS    RESTARTS   AGE
   opendatahub-operator-688bb984bf-59vcn     1/1     Running   0          70s
   $
   ```

   Figure 5.12 – Status of the ODH operator

3. Get the IP address of your minikube environment. This IP address will be used to create ingress for different components of the platform the same way we did for Keycloak. Note that your IP may be different for each minikube instance depending on your underlying infrastructure:

   `minikube ip`

   This command should give you the IP address of your minikube cluster.

4. Open the manifests/kfdef/ml-platform.yaml file and change the value of the following parameters to a NIP (nip.io) domain name of your minikube instance. Only replace the IP address part of the domain name. For example, KEYCLOAK_URL keycloak.<IP Address>.nip.io should become keycloak.192.168.61.72.nip.io. Note that these parameters may be referenced in more than one place in this file. In a full Kubernetes environment, <IP Address> should be the domain name of your Kubernetes cluster:

   I.   KEYCLOAK_URL

   II.   JUPYTERHUB_HOST

III.   `AIRFLOW_HOST`

IV.   `MINIO_HOST`

V.   `MLFLOW_HOST`

VI.   `GRAFANA_HOST`

5. Apply the manifest file to your Kubernetes cluster using the following command:

```
kubectl create -f manifests/kfdef/ml-platform.yaml -n
ml-workshop
```

You should see the following response:

```
$kubectl create -f manifests/kfdef/ml-platform.yaml -n ml-workshop
kfdef.kfdef.apps.kubeflow.org/opendatahub-ml-workshop created
```

Figure 5.13 – Result from applying manifests for the ODH components

6. Start watching the pods being created in the `ml-workshop` namespace by using the following command. It will take a while for all the components to be installed. After several minutes, all the Pods will be in a running state. While the pods are being created, you may see some pods throw errors. This is normal because some pods are dependent on other pods. Be patient as all the components come together and the pods will come into a running state:

```
watch kubectl get pods -n ml-workshop
```

You should see the following response when all the pods are running:

```
Every 2.0s: kubectl get pods -n ml-workshop

NAME                                              READY   STATUS      RESTARTS        AGE
app-aflow-airflow-scheduler-74d8b9c66d-pl6vj      2/2     Running     2 (6m29s ago)   7m51s
app-aflow-airflow-web-c878f7dc9-rm46z             2/2     Running     1 (2m15s ago)   2m33s
app-aflow-airflow-worker-0                        2/2     Running     2 (6m21s ago)   7m51s
app-aflow-postgresql-0                            1/1     Running     0               7m51s
app-aflow-redis-master-0                          1/1     Running     0               7m51s
grafana-76f68bb4c6-9d6mh                          1/1     Running     0               7m51s
jupyterhub-7848ccd4b7-kbd8m                       1/1     Running     0               7m54s
jupyterhub-db-0                                   1/1     Running     0               7m54s
minio-ml-workshop--1-nl544                        0/1     Completed   2               7m51s
minio-ml-workshop-6b84fdc7c4-z7hxk                1/1     Running     0               7m51s
mlflow-579cf69b4d-fqh7r                           2/2     Running     0               7m51s
mlflow-db-0                                       1/1     Running     1 (7m27s ago)   7m51s
prometheus-operator-58cccbff56-wl2dm              1/1     Running     0               7m44s
seldon-controller-manager-7f67f4985b-vxvjt        1/1     Running     0               7m50s
spark-operator-5f78bdd6c6-88tzv                   1/1     Running     0               7m50s
```

Figure 5.14 – CLI response showing the ODH components running on the Kubernetes cluster

So, what does this command do? The **Open Data Hub** (**ODH**) operator consumed the `kfdef` **Custom Resource Definition** (**CRD**) that you have created in *Step 5*. The operator then goes through each of the application objects in the CRD and creates the required Kubernetes objects to run these applications. The Kubernetes objects created in your cluster include several Deployments, Pods, Services, Ingresses, ConfigMaps, Secrets, and PersistentVolumeClaims. You may also run the following command to see all the objects created in the `ml-workshop` namespace:

```
kubectl get all -n ml-workshop
```

You should see all the objects that were created in the `ml-workshop` namespace by the ODH operator.

Congratulations! You have just created a fresh instance of ODH. Now that you have seen the process of creating an instance of the ML platform from a manifest file, it is time to take a look at each of the components of the platform that the data engineers will use for their activities.

Minikube Using Podman Driver

Note that for some `minikube` setups that use `podman` drivers in Linux, the Spark operator may fail due to the limit of the number of threads. To solve this problem, you need to use a kvm2 driver in your `minikube` configuration. You can do this by adding the `--driver=kvm2` parameter to your `minikube start` command.

Understanding and using JupyterHub

Jupyter Notebook has become an extremely popular tool for writing code for ML projects. JupyterHub is a software that facilitates the self-service provisioning of computing environments that includes spinning up pre-configured Jupyter Notebook servers and provisioning the associated compute resources on the Kubernetes platform. On-demand end users such as data engineers and data scientists can provision their own instances of Jupyter Notebook dedicated only to them. If a requesting user already has his/her own running instance of Jupyter Notebook, the hub will just direct the user to the existing instance, avoiding duplicated environments. From the end user's perspective, the whole interaction is seamless. You will see this in the next section of this chapter.

When a user requests an environment in JupyterHub, they are also given the option to choose a pre-configured sizing of hardware resources such as CPU, memory, and storage. This allows for a flexible way for developers, data engineers, and data scientists to provision just the right amount of computing resources for a given task. This dynamic allocation of resources is facilitated by the underlying Kubernetes platform.

Different users may require different frameworks, libraries, and flavors of coding environments. Some data scientists may want to use TensorFlow while others want to use scikit-learn or PyTorch. Some data engineers may prefer to use pandas while some may need to run their data pipelines in PySpark. In JupyterHub, they can configure multiple pre-defined environments for such scenarios. Users can then select a predefined configuration when they request a new environment. These predefined environments are actually container images. This means that the platform operator or platform administrator can prepare several predefined container images that will serve as the end user's computing environment. This feature also enables the standardization of environments. How many times do you have to deal with different versions of the libraries on different developer computers? The standardization of environments can reduce the number of problems related to library version inconsistencies and generally reduce the *it works on my machine* issues.

Figure 5.15 shows the three-step process of provisioning a new JupyterHub environment:

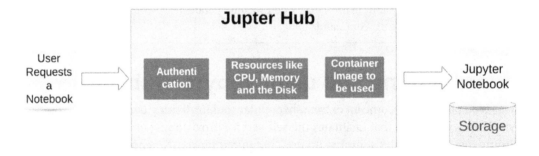

Figure 5.15 – Workflow for creating a new environment in JupyterHub

Now that you know what JupyterHub can do, let's see it in action.

Validating the JupyterHub installation

Every data engineer in the team follows a simple and standard workflow of provisioning an environment. No more manual installations and fiddling with their workstation configurations. This is great for autonomous teams and will definitely help improve your team's velocity.

The ODH operator has already installed the JupyterHub for you in the previous sections. Now, you will spin up a new Jupyter Notebook environment, as a data engineer, and write your data pipelines:

1. Get the ingress objects created in your Kubernetes environment using the following command. We are running this command to find the URL of JupyterHub:

    ```
    kubectl get ingress -n ml-workshop
    ```

    You should see the following example response. Take note of the JupyterHub URL as displayed in the HOSTS column:

```
$kubectl get ingress -n ml-workshop
NAME                   CLASS    HOSTS                              ADDRESS     PORTS      AGE
ap-airflow2            nginx    airflow.192.168.61.72.nip.io       localhost   80, 443    21m
grafana                nginx    grafana.192.168.61.72.nip.io       localhost   80, 443    21m
jupyterhub             nginx    jupyterhub.192.168.61.72.nip.io    localhost   80, 443    21m
minio-ml-workshop-ui   nginx    minio.192.168.61.72.nip.io         localhost   80, 443    21m
mlflow                 nginx    mlflow.192.168.61.72.nip.io        localhost   80, 443    21m
```

Figure 5.16 – All ingresses in your cluster

2. Open a browser from the same machine where minikube is running and navigate to the JupyterHub URL. The URL looks like `https://jupyterhub.<MINIKUBE IP ADDRESS>.nip.io`. This URL will take you to the Keycloak login page to perform SSO authentication. Make sure that you replace the IP address with your minikube IP address in this URL:

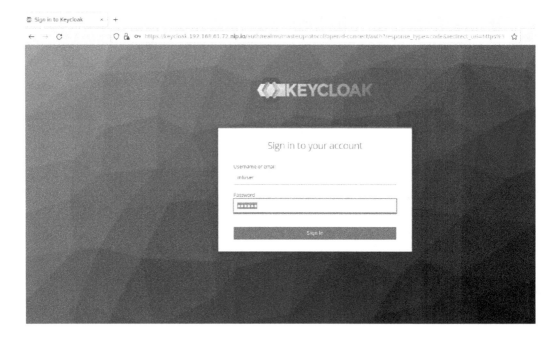

Figure 5.17 – SSO challenge for JupyterHub

3. Type `mluser` for the username, then type whatever password you have set up for this user, and click **Sign In**.

    You will see the landing page of the JupyterHub server where it allows you to select the notebook container image that you want to use and also a predefined size of computing resources you need.

    The notebook image section contains the standard notebooks that you have provisioned using the ODH manifests from the `manifests/jupyterhub-images` folder of the code repository.

    The container size drop-down allows you to select the right size of environment for your need. This configuration is also controlled via the `manifests/jupyterhub/jupyterhub/overlays/mlops/jupyterhub-singleuser-profiles-sizes-configmap.yaml` manifest file.

We encourage you to look into these files to familiarize yourself with what configuration you can set for each manifest.

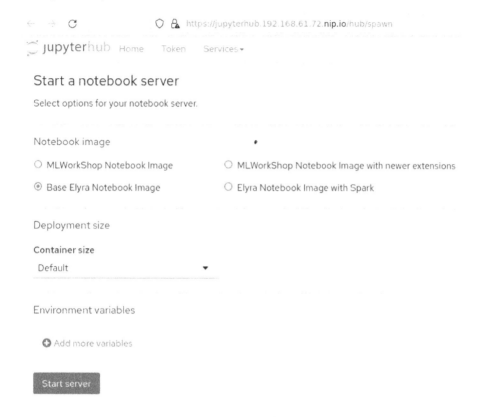

Figure 5.18 – JupyterHub landing page

Select **Base Elyra Notebook Image** and the **Default** container size and hit **Start server**.

4. Validate that a new pod has been created for your user by issuing the following command. Jupyter Notebook instance names start with `jupyter-nb-` and are suffixed with the username of the user. This allows for a unique name of notebook pods for each user:

```
kubectl get pods -n ml-workshop | grep mluser
```

You should see the following response:

Figure 5.19 – Jupyter Notebook pod created by JupyterHub

5. Congratulations! You are now running your own self-provisioned Jupyter Notebook server on the Kubernetes platform.

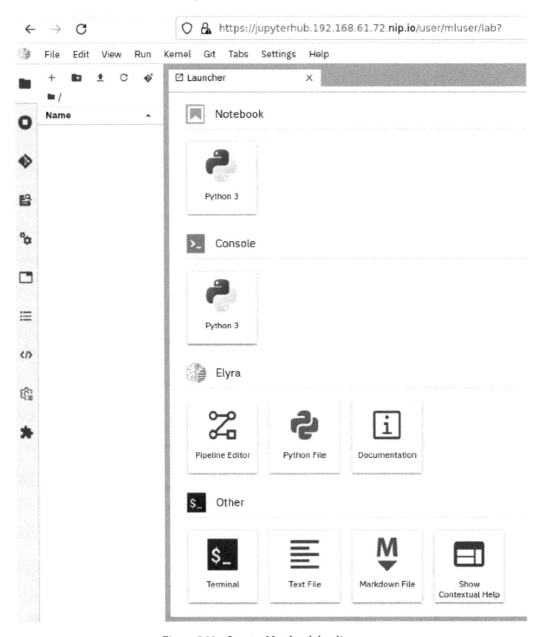

Figure 5.20 – Jupyter Notebook landing page

6.   Now, let's stop the notebook server. Click on the **File > Hub Control Panel** menu option to go to the **Hub Control Panel** page shown as follows:

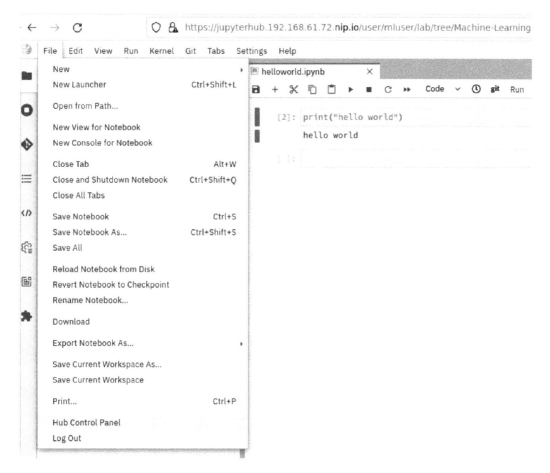

Figure 5.21 – Menu option to see the Hub Control Panel

7. Click on the **Stop My Server** button. This is how you stop your instance of Jupyter Notebook. You may want to start it back again later for the next steps.

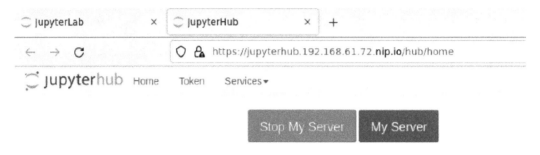

Figure 5.22 – Hub Control Panel

8. Validate that a new pod has been destroyed for your user by issuing the following command:

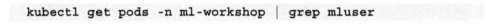

```
kubectl get pods -n ml-workshop | grep mluser
```

There should be no output for this command because the Jupyter Notebook pod has been destroyed by JupyterHub.

We leave it up to you to explore the different bits of the configuration of the notebook in your environment. You will write code using this Jupyter notebook in the later sections of this chapter and the next few chapters of this book, so if you just want to continue reading, you will not miss anything.

Running your first Jupyter notebook

Now that your Jupyter notebook is running, it is time to write the `Hello World!` program. In the code repository of this book, we have provided one such program, and in the following steps, you will check out the code using Git and run the program. Before we start these steps, make sure that you can access your Jupyter notebook using the browser, as mentioned in the preceding section:

1. Click on the Git icon on the left-hand side menu on your Jupyter notebook. The icon is the third from the top. It will display three buttons for different operations. Click on the **Clone a Repository** button:

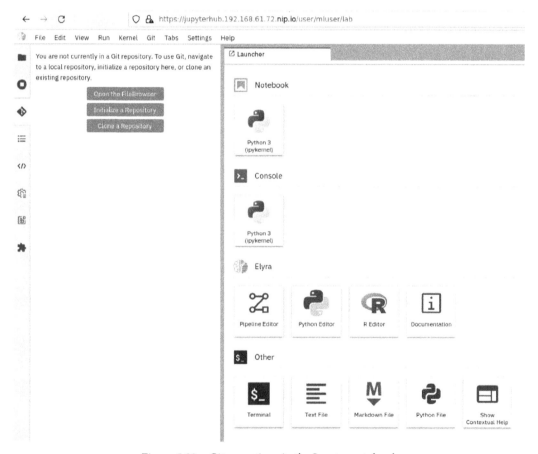

Figure 5.23 – Git operations in the Jupyter notebook

2. In the **Clone a repo** pop-up box, type in the location of the code repository of this book, `https://github.com/PacktPublishing/Machine-Learning-on-Kubernetes.git`, and hit **CLONE**.

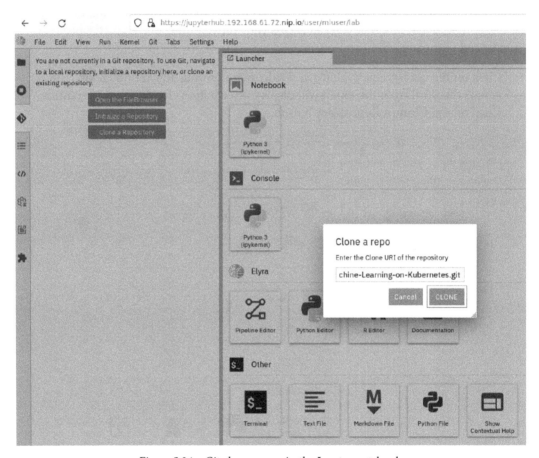

Figure 5.24 – Git clone a repo in the Jupyter notebook

3. You will see that the code repository is cloned onto your Jupyter notebook's file system. As shown in *Figure 5.25*, navigate to the `chapter5/helloworld.ipynb` file and open it in your notebook. Click on the little play icon on the top bar to run the cell:

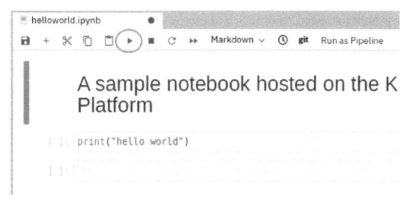

Figure 5.25 – Notebook on your Jupyter environment

4. Et voila! You have just executed a Python code in your own self-provisioned Jupyter Notebook server running on Kubernetes.

5. Shut down your notebook by selecting the **File > Hub Control Panel** menu option. Click on the **Stop My Server** button to shut down your environment. Note that ODH will save your disk and next time you start your notebook, all your saved files will be available.

Congratulations! Now, you can run your code on the platform. Next, we'll get some basics refreshed for the Apache Spark engine.

Understanding the basics of Apache Spark

Apache Spark is an open source data processing engine designed for distributed large-scale processing of data. This means that if you have smaller datasets, say 10s or even a few 100s of GB, a tuned traditional database may provide faster processing times. The main differentiator for Apache Spark is its capability to perform in-memory intermediate computations, which makes Apache Spark much faster than Hadoop MapReduce.

Apache Spark is built for speed, flexibility, and ease of use. Apache Spark offers more than 70 high-level data processing operators that make it easy for data engineers to build data applications, so it is easy to write data processing logic using Apache Spark APIs. Being flexible means that Spark works as a unified data processing engine and works on several types of data workloads such as batch applications, streaming applications, interactive queries, and even ML algorithms.

Figure 5.26 shows the Apache Spark components:

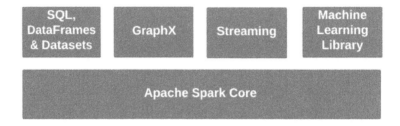

Figure 5.26 – Apache Spark components

Understanding Apache Spark job execution

Most data engineers now know that Apache Spark is a massively parallel data processing engine. It is one of the most successful projects of the Apache Software Foundation. Spark traditionally runs on a cluster of multiple **virtual machines** (**VMs**) or bare metal servers. However, with the popularity of containers and Kubernetes, Spark added support for running Spark clusters on containers on Kubernetes.

There are two most common ways of running Spark on Kubernetes. The first, and the native way, is by using the Kubernetes engine itself to orchestrate the Kubernetes worker pods. In this approach, the Spark cluster instance is always running and the Spark applications are submitted to the Kubernetes API that will schedule the submitted application. We will not dig deeper into how this is implemented. The second approach is through Kubernetes operators. Operators take advantage of the Kubernetes CRDs to create Spark objects natively in Kubernetes. In this approach, the Spark cluster is created on the fly by the Spark operator. Instead of submitting a Spark application to an existing cluster, the operator spins up spark clusters on-demand.

A Spark cluster follows a manager/worker architecture. The Spark cluster manager knows where the workers are located, and the resources available for the worker. The Spark cluster manages the resources for the cluster of worker nodes where your application will run. Each worker has one or more executors that run the assigned jobs through an executor.

Spark applications have two parts, the driver component, and the data processing logic. A driver component is responsible for executing the flow of data processing operations. The driver run first talks to the cluster manager to find out what worker nodes will run the application logic. The driver transforms all the application operations into tasks, schedules them, and assigns tasks directly to the executor processes on the worker node. One executor can run multiple tasks that are associated with the same Spark context.

If your application requires you to collect the computed result and merge them, the driver is the one who will be responsible for this activity. As a data engineer, all this activity is abstracted from you via the SparkSession object. You only need to write the data processing logic. Did we mention Apache Spark aims to be simple?

Figure 5.27 shows the relationship between the Spark driver, Spark cluster manager, and Spark worker nodes:

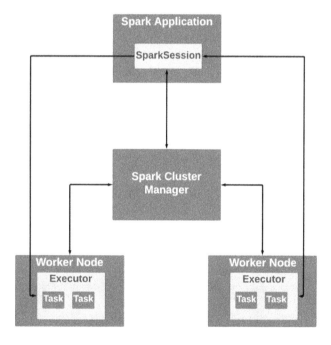

Figure 5.27 – Relationship between Apache Spark components

Understanding how ODH provisions Apache Spark cluster on-demand

We have talked about how the ODH allows you to create a dynamic and flexible development environment to write code such as data pipelines using Jupyter Notebook. We have noticed that data developers need to interact with IT to get time on the data processing clusters such as Apache Spark. These interactions reduce the agility of the team, and this is one of the problems the ML platform solves. To adhere to this scenario, ODH provides the following components:

- A Spark operator that spawns the Apache Spark cluster. For this book, we have forked the original Spark operator provided by ODH and radanalytics to adhere to the latest changes to the Kubernetes API.

- A capability in JupyterHub to issue a request for a new Spark cluster to the Spark operator when certain notebook environments are created by the user.

As a data engineer, when you spin up a new notebook environment using certain notebook images, JupyterHub not only spawns a new notebook server, it also creates the Apache Spark cluster dedicated for you through the Spark operator.

Creating a Spark cluster

Let's first see how the Spark operator works on the Kubernetes cluster. ODH creates the Spark controller. You can see the configuration in the `chapter5/ml-platform.yaml` file under the name `radanalyticsio-spark-cluster`, as shown in *Figure 5.28*. You can see this is another set of Kubernetes YAML files that defines the **custom resource definitions** (**CRDs**), required roles, and the Spark operator deployment. All these files are in the `manifests/radanalyticsio` folder in the code repository of this book.

```
- kustomizeConfig:
    repoRef:
      name: manifests
      path: manifests/radanalyticsio/spark/cluster
  name: radanalyticsio-spark-cluster
```

Figure 5.28 – Snippet of the section of the manifest that installs the Spark operator

When you need to spin up an Apache Spark cluster, you can do this by creating a Kubernetes custom resource called **SparkCluster**. Upon receiving the request, the Spark operator will provision a new Spark cluster as per the required configuration. The following steps will show you the steps for provisioning a Spark cluster on your platform:

1. Validate that the Spark operator pod is running:

    ```
    kubectl get pods -n ml-workshop | grep spark-operator
    ```

    You should see the following response:

Figure 5.29 – Spark operator pod

2. Create a simple Spark cluster with one worker node using the file available at `chapter5/simple-spark-cluster.yaml`. You can see that this file is requesting a Spark cluster with one master and one worker node. Through this custom resource, you can set several Spark configurations, as we shall see in the next section:

```
chapter5 > ! simple-spark-cluster.yaml > {} spec >
1    apiVersion: radanalytics.io/v1
2    kind: SparkCluster
3    metadata:
4      name: simple-spark-cluster
5      namespace: ml-workshop
6    spec:
7      master:
8        instances: '1'
9      metrics: true
10     worker:
11       instances: '1'
```

Figure 5.30 – Spark custom resource

Create this Spark cluster custom resource in your Kubernetes cluster by running the following command. The Spark operator constantly scans for this resource in the Kubernetes platform and automatically creates a new instance of Apache Spark cluster for each given Spark cluster custom resource:

```
kubectl create -f chapter5/simple-spark-cluster.yaml -n
ml-workshop
```

You should see the following response:

```
$kubectl create -f chapter5/simple-spark-cluster.yaml -n ml-workshop
sparkcluster.radanalytics.io/simple-spark-cluster created
```

Figure 5.31 – Response to creating a Spark cluster

3. Validate that the Spark cluster pods are running in your cluster:

```
kubectl get pods -n ml-workshop | grep simple-spark
```

You should see the following response. There are two pods created by the Spark operator, one for the Spark master node and another for the worker node. The number of worker pods depends on the value of the `instances` parameters in the `SparkCluster` resource. It may take some time for the pods to come to a running state the first time:

```
$kubectl get pods -n ml-workshop | grep simple-spark
simple-spark-cluster-m-98hl7                              1/1      Running
simple-spark-cluster-w-jwntc                              1/1      Running
$
```

Figure 5.32 – List of running Spark cluster pods

Now, you know how the Spark operator works on the Kubernetes cluster. The next step is to see how JupyterHub is configured to request the cluster dynamically while provisioning a new notebook for you.

Understanding how JupyterHub creates a Spark cluster

Simply put, JupyterHub does what you did in the preceding section. JupyterHub creates a `SparkCluster` resource in Kubernetes so that the Spark operator can provision the Apache Spark cluster for your use. This `SparkCluster` resource configuration is a Kubernetes `ConfigMap` file and can be found at `manifests/jupyterhub/jupyterhub/base/jupyterhub-spark-operator-configmap.yaml`. Look for `sparkClusterTemplate` in this file, as shown in *Figure 5.33*. You can see that it looks like the file you have created in the previous section:

```
manifests > jupyterhub > jupyterhub > base >  !  jupyterhub-spark-operator-configmap.yaml >
20       sparkClusterTemplate: |
21         kind: SparkCluster
22         apiVersion: radanalytics.io/v1
23         metadata:
24           name: "spark-cluster-{{ user }}"
25         spec:
26           worker:
27             instances: "{{ worker_nodes }}"
28             memoryLimit: "{{ worker_memory_limit }}"
29             cpuLimit: "{{ worker_cpu_limit }}"
30             memoryRequest: "{{ worker_memory_request }}"
31             cpuRequest: "{{ worker_cpu_request }}"          You, 3 weeks ago •
32           master:
33             instances: "{{ master_nodes }}"
34             memoryLimit: "{{ master_memory_limit }}"
35             cpuLimit: "{{ master_cpu_limit }}"
36             memoryRequest: "{{ master_memory_request }}"
37             cpuRequest: "{{ master_cpu_request }}"
38           customImage: "{{ spark_image }}"
39           metrics: true
40           env:
41           - name: SPARK_METRICS_ON
42             value: prometheus
```

Figure 5.33 – JupyterHub template for Spark resources

Some of you might have noticed that this is a template, and it needs the values for specific variables mentioned in this template. Variables such as {{ user }} and {{ worker_ nodes }} and so on. Recall that we have mentioned that JupyterHub creates the SparkCluster request while it is provisioning a container for your notebook. JupyterHub uses this file as a template and fills in the values while creating your notebook. How does JupyterHub decide to create a Spark cluster? This configuration is called **profiles** and is available as a ConfigMap file in manifests/jupyterhub/jupyterhub/overlays/ spark3/jupyterhub-singleuser-profiles-configmap.yaml. This looks like the file shown in *Figure 5.33*.

You can see that the `image` field specifies the name of the container image on which this profile will be triggered. So, as a data engineer, when you select this notebook image from the JupyterHub landing page, JupyterHub will apply this profile. The second thing in the profile is the `env` section, which specifies the environment variables that will be pushed to the notebook container instance. The `configuration` object defines the values that will be applied to the template that is mentioned in the `resources` key:

```
- name: Spark Notebook
  images:
  - quay.io/ml-aml-workshop/elyra-spark:0.0.4
  env:
    - name: PYSPARK_SUBMIT_ARGS
      value: '--conf spark.cores.max=2 --conf spark.executor.instances=1 --conf spark.exec
    - name: PYSPARK_DRIVER_PYTHON
      value: 'jupyter'
    - name: PYSPARK_DRIVER_PYTHON_OPTS
      value: 'notebook'
    - name: PYTHONPATH
      value: '$PYTHONPATH:/opt/app-root/lib/python3.8/site-packages/:/opt/app-root/lib/pyt


  services:
    spark:
      resources:
      - name: spark-cluster-template
        path: notebookPodServiceTemplate
      - name: spark-cluster-template
        path: sparkClusterTemplate
      configuration:
        worker_nodes: '1'
        master_nodes: '1'
        master_memory_limit: '2Gi'
        master_cpu_limit: '750m'
        master_memory_request: '2Gi'
        master_cpu_request: '100m'
        worker_memory_limit: '2Gi'
        worker_cpu_limit: '750m'
        worker_memory_request: '2Gi'        You, a week ago • move images to a central repo
        worker_cpu_request: '250m'
        spark_image: 'quay.io/ml-on-k8s/spark:3.0.0'
      return:
        SPARK_CLUSTER: 'metadata.name'
```

Figure 5.34 – JupyterHub profile for Spark resources

As you may appreciate, there is a lot of work done behind the scenes to make a streamlined experience for you and your team, and in the true sense of open source, you can configure everything and even give back to the project if you come up with any modifications or new features.

In the next section, you will see how easy it is to write and run a Spark application on your platform running these components.

Writing and running a Spark application from Jupyter Notebook

Before you run the following steps, make sure that you grasped the components and their interactions that we have introduced in the previous section of this chapter:

1. Validate that the Spark operator pod is running by running the following command:

    ```
    kubectl get pods -n ml-workshop | grep spark-operator
    ```

    You should see the following response:

Figure 5.35 – Spark operator pod

2. Validate that the JupyterHub pod is running by running the following command:

    ```
    kubectl get pods -n ml-workshop | grep jupyterhub
    ```

    You should see the following response:

```
$kubectl get pods -n ml-workshop | grep jupyterhub
jupyterhub-7848ccd4b7-kbd8m                    1/1      Running
jupyterhub-db-0                                1/1      Running
```

Figure 5.36 – JupyterHub pod

3. Before you start the notebook, let's delete the Spark cluster you have created in the previous sections by running the following command. This is to demonstrate that JupyterHub will automatically create a new instance of Spark cluster for you:

    ```
    kubectl delete sparkcluster simple-spark-cluster -n
    ml-workshop
    ```

4. Log in to your JupyterHub server. Refer to the *Validating JupyterHub configuration* section earlier in this chapter. You will get the landing page of your server. Select the **Elyra Notebook Image with Spark** image and the **Small** container size. This is the same image that you have configured in the `manifests/jupyterhub/jupyterhub/overlays/spark3/jupyterhub-singleuser-profiles-configmap.yaml` file.

5. Click on **Start server**:

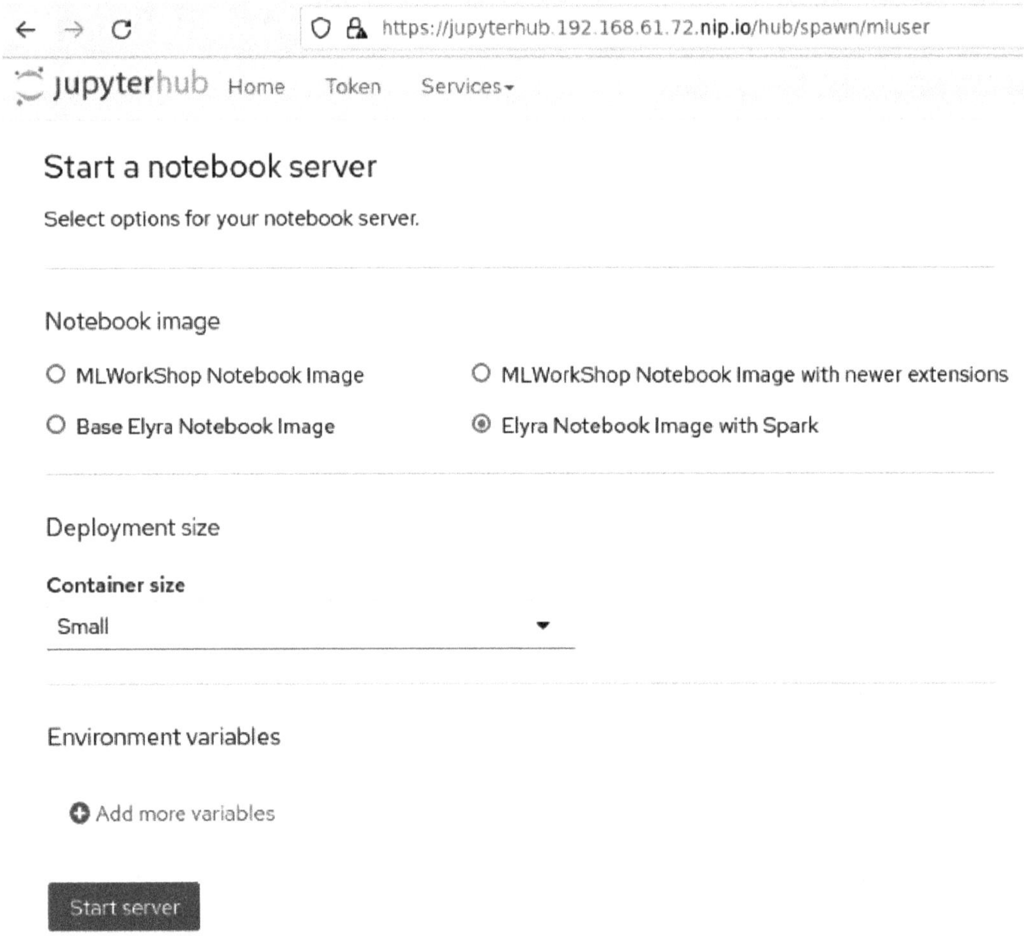

Figure 5.37 – JupyterHub landing page showing Elyra Notebook Image with Spark

The notebook you have just started will also trigger the creation of a dedicated Spark cluster for you. It may take some time for the notebook to start because it has to wait for the Spark cluster to be ready.

Also, you may have noticed that the image you have configured in the `jupyterhub-singleuser-profiles-configmap.yaml` file is `quay.io/ml-aml-workshop/elyra-spark:0.0.4`, while the name we have selected is **Elyra Notebook Image with Spark**, and they are not the same. The mapping of the image with the descriptive text is configured in the `manifests/jupyterhub-images/elyra-notebook-spark3-imagestream.yaml` file. You will find the descriptive text displayed on the JupyterHub landing page coming from the *annotations* section of this file. If you want to add your own images with specific libraries, you can just add another file here and it will be available for your team. This feature of JupyterHub enables the standardization of notebook container images, which allows everyone in the teams to have the same environment configurations and the same set of libraries.

6. After the notebook has started, validate that the Spark cluster is provisioned for you. Note that this is the Spark cluster for the user of this notebook and is dedicated to this user only:

```
kubectl get pods -n ml-workshop | grep mluser
```

You should see the following response. The response contains a notebook pod and two Spark pods; the one with a little `-m` character is the master, while the other is the worker. Notice how your username (`mluser`) is associated with the pod names:

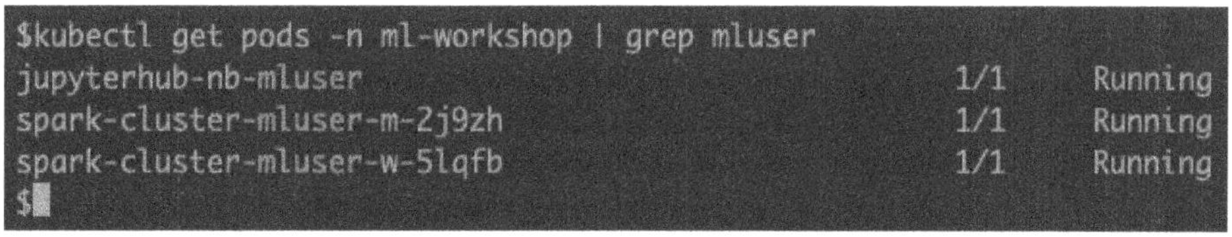

```
$kubectl get pods -n ml-workshop | grep mluser
jupyterhub-nb-mluser                      1/1     Running
spark-cluster-mluser-m-2j9zh              1/1     Running
spark-cluster-mluser-w-5lqfb              1/1     Running
$
```

Figure 5.38 – Jupyter Notebook and Spark cluster pods

Now, everyone in your team will get their own developer environment with dedicated Spark instances to write and test the data processing code.

7. Apache Spark provides a UI through which you can monitor applications and data processing jobs. The ODH-provisioned Spark cluster provides this GUI, and it is available at `https://spark-cluster-mluser.192.168.61.72.nip.io`. Make sure to change the IP address to your minikube IP address. You may also notice that the username you have used to log in to JupyterHub, `mluser`, is part of the URL. If you have used a different username, you may need to adjust the URL accordingly.

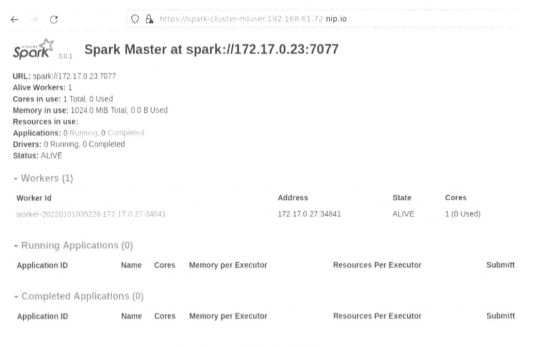

Figure 5.39 – Spark UI

The preceding UI mentions that you have one worker in the cluster, and you can click on the worker node to find out the executors running inside the worker node. If you want to refresh your knowledge of the Spark cluster, please refer to the *Understanding the basics of Apache Spark* section earlier in this chapter.

8. Open the `chapter5/hellospark.ipynb` file from your notebook. This is quite a simple job that calculates the square of the given array. Remember that Spark will automatically schedule the job and distribute it among executors. The notebook here is the Spark Driver program, which talks to the Spark cluster, and all of this is abstracted via the `SparkSession` object.

On the second code cell of this notebook, you are creating a `SparkSession` object. The `getOrCreateSparkSession` utility function will connect to the Spark cluster provisioned for you by the platform.

The last cell is where your data processing logic resides. In this example, the logic was to take the data and calculate the square of each element in the array. Once the data is processed, the `collect` method will bring the response to the driver that is running in the Spark application in your notebook.

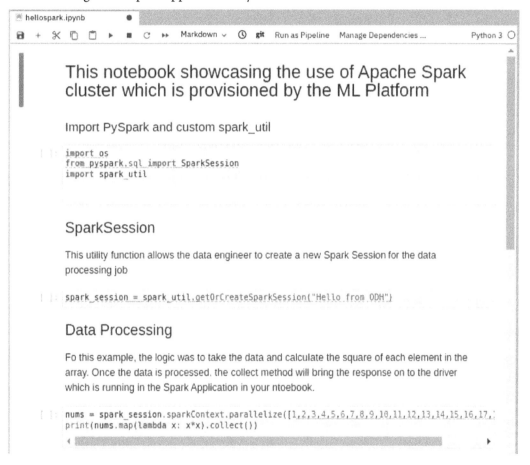

Figure 5.40 – A notebook with a simple Spark application

Click on the **Run > Run All cells** menu option, and the notebook will connect to the Spark cluster, and submit and execute your job.

9. While the job is progressing, open the Spark UI at `https://spark-cluster-mluser.192.168.61.72.nip.io`. Remember to adjust the IP address as per your settings, and click on the table with the **Application ID** heading under the **Running Applications** heading on this page.

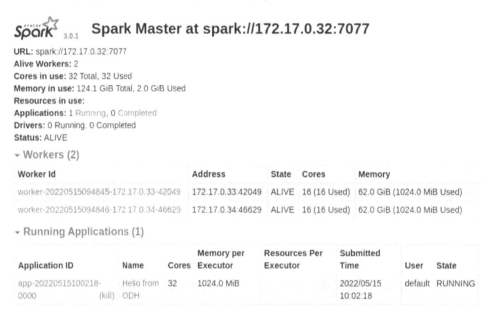

Figure 5.41 – Apache Spark UI

10. Navigate to the details page of the Spark application. Note that the application title, **Hello from ODH**, has been set up in your notebook. Click on the **Application Detail UI** link:

Figure 5.42 – Spark UI showing the submitted Spark job

You should see a page showing the detailed metrics of the job that you have just executed on the Spark cluster from your Jupyter notebook:

Figure 5.43 – Spark UI showing the submitted job details

11. Once you are done with your work, go to the **File > Hub Control Panel** menu option and click on the **Stop My Server** button:

Figure 5.44 – Jupyter Notebook control panel

12. Validate that the Spark cluster has been terminated by issuing the following command:

```
kubectl get pods -n ml-workshop | grep mluser
```

You should not see a response because the pods are terminated by the Spark operator on your cluster.

You have finally run a basic data processing job in an on-demand ephemeral Spark cluster that is running on Kubernetes. Note that you have done all this from a Jupyter notebook running on Kubernetes.

With this capability in the platform, data engineers can perform huge data processing tasks directly from the browser. This capability also allows them to collaborate easily with each other to provide transformed, cleaned, high-quality data for your ML project.

Summary

In this chapter, you have just created your first ML platform. You have configured the ODH components via the ODH Kubernetes operator. You have seen how a data engineer persona will use JupyterHub to provision the Jupyter notebook and the Apache Spark cluster instance while the platform provides the provisioning of the environments automatically. You have also seen how the platform enables standardization of the operating environment via the container images, which bring consistency and security. You have seen how a data engineer could run Apache Spark jobs from the Jupyter notebook.

All these capabilities allow the data engineer to work autonomously and in a self-serving fashion. You have seen that all these components were available autonomously and on-demand. The elastic and self-serving nature of the platform will allow teams to be more productive and agile while responding to the ever-changing requirements of the data and the ML world.

In the next chapter, you will see how data scientists can benefit from the platform and be more efficient.

6
Machine Learning Engineering

In this chapter, we will move the discussion to the model building and model management activities of the **machine learning** (**ML**) engineering lifecycle. You will learn about the ML platform's role of providing a self-serving solution to data scientist so they can work more efficiently and collaborate with data teams and fellow data scientists.

The focus of this chapter is not on building models; instead, it is on showing how the platform can bring consistency and security across different environments and different members of your teams. You will learn how the platform simplifies the work of data scientists in terms of preparing and maintaining their data science workspaces.

In this chapter, you will learn about the following topics:

- Understanding ML engineering?
- Using a custom notebook image
- Introducing MLflow
- Using MLflow as an experiment tracking system
- Using MLflow as a model registry system

Technical requirements

This chapter includes some hands-on setup and exercises. You will need a running Kubernetes cluster configured with **Operator Lifecycle Manager** (**OLM**). Building such a Kubernetes environment is covered in *Chapter 3*, *Exploring Kubernetes*. Before attempting the technical exercises in this chapter, please make sure that you have a working Kubernetes cluster and that **Open Data Hub** (**ODH**) is installed on your Kubernetes cluster. Installing ODH is covered in *Chapter 4*, *The Anatomy of a Machine Learning Platform*. You can find all the code associated with this book at `https://github.com/PacktPublishing/Machine-Learning-on-Kubernetes`.

Understanding ML engineering

ML engineering is the process of applying software engineering principles and practices to ML projects. In the context of this book, ML engineering is also a discipline that facilitates applying application development practices to the data science lifecycle. When you write a traditional application such as a website or a banking system, there are processes and tools to assist you in writing high-quality code right from the start. Smart IDEs, standard environments, continuous integration, automated testing, and static code analysis are just a few examples. Automation and continuous deployment practices enable organizations to deploy applications many times in a day and with no downtime.

ML engineering is a loose term that brings the benefits of traditional software engineering practices to the model development world. However, most data scientists are not developers. They may not be familiar with software engineering practices. Also, the tools that the data scientists use may not be the right tools to perform ML engineering tasks. Having said that, the model is just another piece of software. Therefore, we can also apply existing software engineering approaches to ML models. Using containers to package and deploy ML models is one such example.

Some teams may employ ML engineers to supplement the work of data scientists. While the data scientist's primary responsibility is to build ML or deep learning models that solve business problems, ML engineers focus more on the software engineering facets. Some of the responsibilities of data engineers include the following:

- Model optimization (also about making sure that the built model is optimized for the target environment where the model will be hosted).

- Model packaging (making ML models portable, shippable, executable, and version-controlled). Model packaging may also include model serving and containerization.

- Monitoring (establishing an infrastructure for collecting performance metrics, logging, alerting, and anomaly detection such as drift and outlier detection).

- Model testing (including facilitation and automation of A/B testing).

- Model deployment.

- Building and maintenance of MLOps infrastructure.

- Implementation of continuous integration and continuous deployment pipelines for ML models.

- Automation of ML lifecycle processes.

There are other responsibilities of ML engineers that are not listed in the preceding list, but this list should already give you an idea of how to differentiate data science from ML engineering.

The ML platform that you are building will reduce the number of ML engineering tasks to be done manually to a point where even the data scientists can do most of the ML engineering tasks by themselves.

In the next sections, you will see how data scientists can track the model development iterations to improve model quality and share the learning with the team. You will see how teams can apply version control to ML models and other practices of software engineering to the ML world.

We will continue our journey of ML engineering into the next chapter, where you will see how models can be packaged and deployed in a standard way and see how the deployment process can be automated.

Let's start with building standard development environments for our data science team.

Using a custom notebook image

As you have seen in *Chapter 5*, *Data Engineering*, JupyterHub allows you to spin up Jupyter Notebook-based development environments in a self-service manner. You have launched the **Base Elyra Notebook Image** container image and used it to write the data processing code using Apache Spark. This approach enables your team to use a consistent or standardized development environment (for example, same Python versions and same libraries for building code) and apply security policies to the known set of software being used by your team. However, you may also want to create your own custom images with a different set of libraries or a different ML framework. The platform allows you to do that.

In the following subsection, you will build and deploy a custom container image to be used within your team.

Building a custom notebook container image

Let's assume that your team wants to use a specific version of the Scikit library along with some other supporting libraries such as `joblib`. You then want your team to use this library while developing their data science code:

1. Open the `Dockerfile` provided in the code repository of this book at `chapter6/CustomNotebookDockerfile`. This file uses the base image provided and used by ODH and then adds the required libraries. The file is shown in *Figure 6.1*:

```
chapter6 >  CustomNotebookDockerfile > ...
  1    FROM quay.io/thoth-station/s2i-lab-elyra:v0.1.1
  2
  3    # copy the requirements.txt file whoch contains all the deps
  4    COPY requirements.txt requirements.txt
  5    RUN pip install -r requirements.txt
  6
  7    # you can install dependices directly too
  8    RUN pip install watermark
```

```
 requirements.txt U X

chapter6 >  requirements.txt
  1    scikit-learn==1.0
  2    joblib==1.1.0
  3
```

Figure 6.1 – Dockerfile for the custom notebook image

Note the first line, which refers to the latest image at the time of writing. This image is used by ODH. Lines 4 and 5 install the Python packages defined in the `requirements.txt` file. Line 8 installs the dependencies that are not in the `requirements.txt` file. If you wish to add additional packages to the image, you can simply insert a line in `requirements.txt`.

2. Build the image using the file provided in the preceding step. Run the following command:

```
docker build -t scikit-notebook:v1.1.0 -f chapter6/
CustomNotebookDockerfile ./chapter6/.
```

You should see the following response:

```
=> [2/4] COPY requirements.txt requirements.txt
=> [3/4] RUN pip install -r requirements.txt
=> [4/4] RUN pip install watermark
=> exporting to image
=> => exporting layers
=> => writing image sha256:d6efdbc94979fb2caa23c08e9a8599a2e8159bf8a19e9aa602d2123f7685d408
=> => naming to docker.io/library/scikit-notebook:v1.1.0

Use 'docker scan' to run Snyk tests against images to find vulnerabilities and learn how to fix them
```

Figure 6.2 – Output of the container build command

3. Tag the built image as per your liking. You will need to push this image to a registry from where your Kubernetes cluster can access it. We use quay.io as the public Docker repository of choice, and you can use your preferred repository here. Notice that you will need to adjust the following command and change the quay.io/ml-on-k8s/ part before execution of the command:

```
docker tag scikit-notebook:v1.1.0 quay.io/ml-on-k8s/
scikit-notebook:v1.1.0
```

There is no output of the preceding command.

4. Push the image to the Docker repository of your choice. Use the following command and make sure to change the repository location as per *Step 3*. This image may take some time to be pushed to an internet repository based on your connection speed. Be patient:

```
docker push quay.io/ml-on-k8s/scikit-notebook:v1.1.0
```

You should see the output of this command as shown in *Figure 6.3*. Wait for the push to complete.

```
$docker push quay.io/ml-on-k8s/scikit-notebook:v1.1.0
The push refers to repository [quay.io/ml-on-k8s/scikit-notebook]
d67593a5dd34: Pushed
ff0c8535b565: Pushing [===>                                           ]  20.07MB/309.2MB
d3e12cc0c2ad: Pushed
2c26d8e0f927: Mounted from thoth-station/s2i-lab-elyra
c5e4d2689426: Mounted from thoth-station/s2i-lab-elyra
6428ea2f8e03: Mounted from thoth-station/s2i-lab-elyra
756159f326e3: Mounted from thoth-station/s2i-lab-elyra
b65b0a6ccebd: Mounted from thoth-station/s2i-lab-elyra
e5e0593b8dea: Waiting
8acf87721326: Waiting
37a9b7fa894e: Waiting
b093a0249a2a: Waiting
08595b8cc66c: Waiting
13c00070aba8: Waiting
743f2ab3a4c4: Waiting
cfc084379a4f: Waiting
714b6d92c03a: Waiting
168d524b4a7c: Waiting
0488bd866f64: Waiting
0d3f22d60daf: Waiting
```

Figure 6.3 – Pushing the custom notebook image to a Docker repository

Now, the image is available to be used. You will configure ODH manifests in the next steps to use this image.

5. Open the `manifests/jupyterhub-images/base/customnotebook-imagestream.yaml` file. This file is shown as follows:

```
manifests > jupyterhub-images > base > ! customnotebook-imagestream.yaml > {} status > [ ] tags > {}
1    apiVersion: image.openshift.io/v1
2    kind: ImageStream
3    metadata:
4      labels:
5        opendatahub.io/notebook-image: "true"
6      annotations:
7        opendatahub.io/notebook-image-name: "SciKit v1.10 - Elyra Notebook Image"
8        opendatahub.io/notebook-image-desc: "Jupyter notebook image with Scikit"
9      name: elyra-scikit-customnotebook
10   spec:
11     tags:
12       from:
13         kind: DockerImage
14         name: quay.io/ml-on-k8s/scikit-notebook:v1.1.0
15       name: latest
16   status:
17     tags:
18       - tag: latest
19         dockerImageReference: quay.io/ml-on-k8s/scikit-notebook:v1.1.0
```

Figure 6.4 – ImageStream object

JupyterHub from ODH uses a CRD called **Imagestream**. This is a native object on Red Hat OpenShift, but it is not available in standard Kubernetes. We have created this object as a custom resource in the manifests of ODH so that it can integrate with upstream Kubernetes. You can find these resources at `manifests/odh-common/base/imagestream-crd.yaml`.

Notice on lines 7 and 8, we have defined some annotations. JupyterHub reads all the `imagestream` objects and uses these annotations to be displayed on the JupyterHub landing page. JupyterHub also looks at the field named `dockerImageReference` to load these container images upon request.

We encourage you to fork the code repository of this book onto your own Git account and add more images. Keep in mind to change the location of the Git repository in the `manifests/kfdef/ml-platform.yaml` file.

6. For the JupyterHub server to see the newly created image, you will need to restart the JupyterHub pod. You can find the pod via the following command and delete this pod. After a few seconds, Kubernetes will restart this pod and your new image will appear on the JupyterHub landing page:

```
kubectl get pods -n ml-workshop | grep jupyterhub
```

You should see the following response. Note that the pod name will be different for your setup:

```
$kubectl get pods -n ml-workshop | grep jupyterhub
jupyterhub-7848ccd4b7-thnmm                              1/1     Running
jupyterhub-db-0                                          1/1     Running
```

Figure 6.5 – Pods with names containing jupyterhub

7. Delete the JupyterHub pod by running the following command. Note that you do not need to delete this pod for this exercise, because the custom image is already present in our manifest files. This step will be required once you add a new customer notebook image using the steps mentioned in this section:

```
kubectl delete pod jupyterhub-7848ccd4b7-thnmm -n
ml-workshop
```

You should see the following response. Note that the pod name will be different for your setup:

```
$kubectl delete pod jupyterhub-7848ccd4b7-thnmm -n ml-workshop
pod "jupyterhub-7848ccd4b7-thnmm" deleted
```

Figure 6.6 – Output of the delete pod command

8. Log in to JupyterHub and you will see the new notebook image listed there:

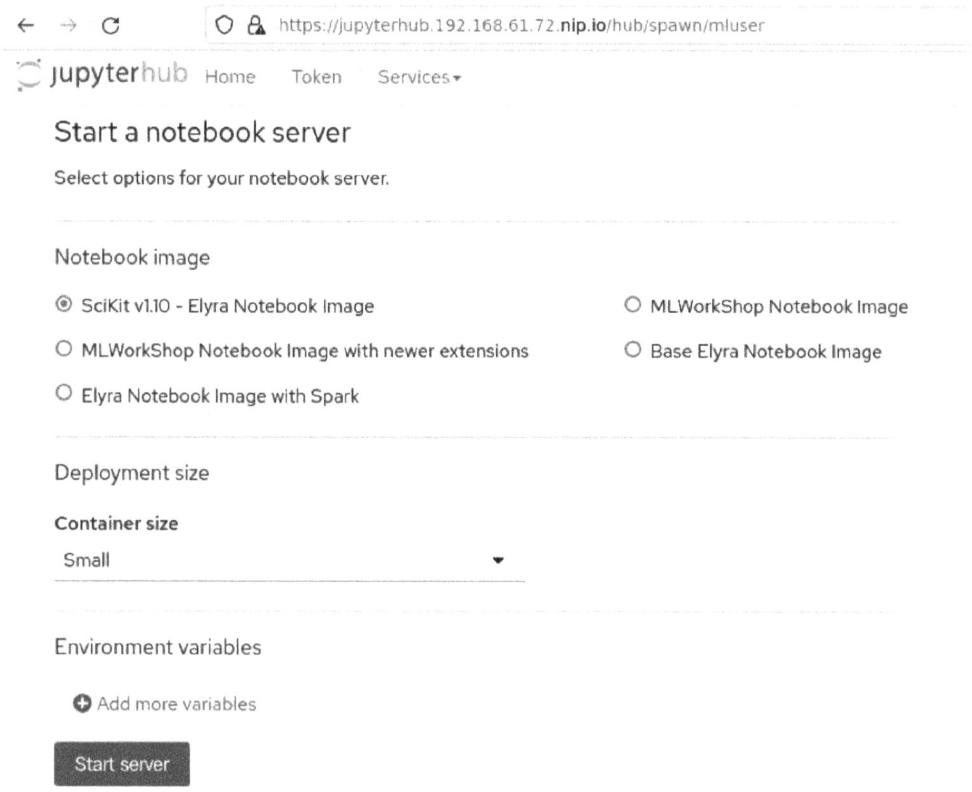

Figure 6.7 – JupyterHub landing page showing the new notebook image

In the next section, you will learn about MLflow, a software that assists teams in recording and sharing the outcomes of model training and tuning experiments.

Introducing MLflow

Simply put, MLflow is there to simplify the model development lifecycle. A lot of the data scientist's time is spent finding the right algorithms with the right hyperparameters for the given dataset. As a data scientist, you experiment with different combinations of parameters and algorithms, then review and compare the results to make the right choice. MLflow allows you to record, track, and compare these parameters, their results, and associated metrics. The component of MLflow that captures the details of each of your experiments is called the **tracking server**. The tracking server captures the environment details of your notebook, such as the Python libraries and their versions, and the artifacts generated by your experiment.

The tracking server allows you to compare the data captured between different runs of an experiment, such as the performance metrics (for example, accuracy) alongside the hyperparameters used. You can also share this data with your team for collaboration.

The second key capability of the MLflow tracking server is the model registry. Consider that you have run ten different experiments for the given dataset, while each of the experiments resulted in a model. Only one of the models will be used for the given problem. The model registry allows you to tag the selected model with one of the three stages (**Staging**, **Production**, and **Archived**). The model registry has APIs that allow you to access these models from your automated jobs. Versioning models in a registry will enable you to roll back to previous versions of the model using your automation tools in production if needed.

Figure 6.8 shows the two major capabilities of the MLflow software:

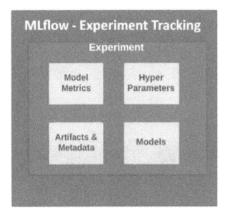 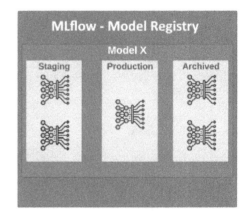

Figure 6.8 – MLflow major capabilities

Now that you know what MLFlow is used for, let's take a look at the components that made up MLFlow.

Understanding MLflow components

Let's see what the major components of the MLflow system are and how it fits into our ML platform's ecosystem.

MLflow server

MLflow is deployed as a container, and it contains a backend server, a GUI, and an API to interact with it. In the later sections of this chapter, you will use the MLflow API to store the experiment data onto it. You will use the GUI component to visualize experiment tracking and the model registry parts. You can find this configuration at `manifests/mlflow/base/mlflow-dc.yaml`.

MLflow backend store

The MLflow server needs a backend store to store the metadata about experiments. The ODH component automatically provisions a PostgreSQL database to be used as a backend store for MLflow. You can find this configuration at `manifests/mlflow/base/mlflow-postgres-statefulset.yaml`.

MLflow storage

The MLflow server supports several types of storage, such as S3 and databases. This storage will serve as the persistent storage for the artifacts, such as files and model files. In our platform, you will provision an open source S3 compatible storage service known as **Minio**. Minio will provide the S3 API capabilities to the platform; however, your organization may already have an enterprise-wide S3 solution, and we recommend using the existing solution if there is one. You can find this configuration at `manifests/minio/base/minio-dc.yaml`.

MLflow authentication

MLflow does not have an out-of-the-box authentication system at the time of writing. In our platform, we have configured a proxy server in front of the MLflow GUI that will authenticate the request before forwarding it to the MLflow server. We are using the open source component at `https://github.com/oauth2-proxy/oauth2-proxy` for this purpose. The proxy has been configured to perform **Single-Sign-On (SSO)** with the **Keycloak** service of the platform.

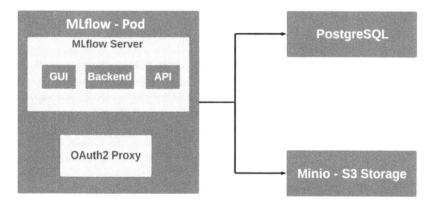

Figure 6.9 – MLflow and associated components in the platform

As you can see in *Figure 6.9*, the MLflow pod has two containers in it: the MLflow server and the OAuth2 proxy. The Oauth2 proxy has been configured to use the Keycloak instance you installed.

When you created a new instance of ODH in *Chapter 5*, *Data Engineering*, it installed many platform components, including MLflow and Minio. Now, let's validate the MLflow installation.

Validating the MLflow installation

ODH has already installed the MLflow and associated components for you. Now, you will use the MLflow GUI to get yourself familiar with the tool. You can imagine all the team members will have access to experiments and models, which will improve your team's collaboration:

1. Get the ingress objects created in your Kubernetes environment using the following command. This is to get the URL of the endpoints where our services are deployed:

    ```
    kubectl get ingress -n ml-workshop
    ```

    You should see the following response:

    ```
    $kubectl get ingress -n ml-workshop
    NAME                    CLASS   HOSTS                                 ADDRESS     PORTS     AGE
    ap-airflow2             nginx   airflow.192.168.61.72.nip.io          localhost   80, 443   21m
    grafana                 nginx   grafana.192.168.61.72.nip.io          localhost   80, 443   21m
    jupyterhub              nginx   jupyterhub.192.168.61.72.nip.io       localhost   80, 443   21m
    minio-ml-workshop-ui    nginx   minio.192.168.61.72.nip.io            localhost   80, 443   21m
    mlflow                  nginx   mlflow.192.168.61.72.nip.io           localhost   80, 443   21m
    ```

    Figure 6.10 – All ingress objects in your cluster namespace

2. Open the Minio GUI, our S3 component, and validate that there is a bucket available for MLflow to be used as its storage. The URL for the Minio component will look like `https://minio.192.168.61.72.nip.io`, where you will adjust the IP address as per your environment. The password is configured in the manifests file, and it is `minio123`. We have added Minio to the manifests to show that there is an option available using open source technologies, but making it suitable for production is out of scope for this book. Click on the buckets menu item on the left-hand side of the screen and you will see the available buckets:

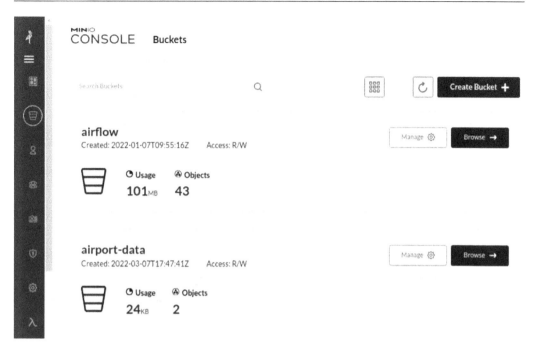

Figure 6.11 – Minio bucket list

How are all of these buckets created? In the manifests, we have a Kubernetes job that creates the buckets. You can find the job at `manifests/minio/base/minio-job.yaml`. The job is using the Minio command-line client, `mc`, to create the buckets. You can find these commands under the `command` field name in this file.

The configuration of S3 that is being used by MLflow is configured at `manifests/mlflow/base/mlflow-dc.yaml file`.

You can see the settings as follows:

```
manifests > mlflow > base >  !  mlflow-dc.yaml >     apiVersion
60              env:
61                 - name: MLFLOW_S3_ENDPOINT_URL
62                   value: http://minio-ml-workshop:9000/
63                 - name: AWS_ACCESS_KEY_ID
64                   value: "minio"
65                 - name: AWS_SECRET_ACCESS_KEY
66                   valueFrom:
67                     secretKeyRef:
68                       name: mlflow-store
69                       key: AWS_SECRET_ACCESS_KEY
70                 - name: AWS_BUCKET_NAME
71                   value: "mlflow"
72                 - name: MLFLOW_S3_IGNORE_TLS
73                   value: "true"
74              ports:
```

Figure 6.12 – MLflow configuration to use Minio

3. Open a browser and paste the HOSTS value for the jupyterhub ingress into your browser. For me, it was https://mlflow.192.168.61.72.nip.io. This URL will take you to the Keycloak login page, which is the SSO server as shown in the following figure. Make sure that you replace the IP address with yours in this URL. Recall that the authentication part of MLflow is being managed by a proxy that you have configured in manifests/mlflow/base/mlflow-dc.yaml.

4. You can see the configuration of the OAuth proxy for MLflow as follows. Because oauth-proxy and MLflow belong to the same pod, all we have done is route the traffic from oauth-proxy to the MLflow container. This is set up with the -upstream property. You can also see oauth-proxy needs the name of the identity provider server, which is Keycloak, and it is configured under the -oidc-issuer property:

```
manifests > mlflow > base > ! mlflow-dc.yaml > {} spec > {} template > {} spec > [ ] containers >
22          containers:
23            - name: oauth-proxy
24              image: 'quay.io/oauth2-proxy/oauth2-proxy:v7.2.0'
25              args:
26                - '--provider=keycloak-oidc'
27                - '--https-address='
28                - '--http-address=:5000'
29                - '--client-id=mlflow'
30                - '--client-secret=ad216993-cf3a-4742-ba17-b531d5c22046'
31                - '--upstream=http://localhost:5500'
32                - '--email-domain=*'
33                - '--cookie-secret=ad12ca-qw23asd55adcwbjep'
34                - '--oidc-issuer-url=https://$(KEYCLOAK_URL)/auth/realms/master'
35                - '--allowed-role=mflow:admin'
36                - --reverse-proxy=true
37                - --skip-provider-button=true
38                - --ssl-insecure-skip-verify=true
39                - --ssl-upstream-insecure-skip-verify=true
40              ports:
41                - name: public
42                  containerPort: 5000
43                  protocol: TCP
44              resources: {}
45              terminationMessagePath: /dev/termination-log
46              terminationMessagePolicy: File
47              imagePullPolicy: IfNotPresent
48
```

Figure 6.13 – OAuth proxy configuration for MLflow

The landing page of MLflow looks like the page in *Figure 6.14*. You will notice there are two sections on the top bar menu. One has the label **Experiments** and the other one, **Models**.

5. Before you see this page, the SSO configuration will display the login page. Enter the user ID as `mluser` and the password as `mluser` to log in. The username and password were configured in *Chapter 4, The Anatomy of a Machine Learning Platform*, in the *Creating a Keycloak user* section.

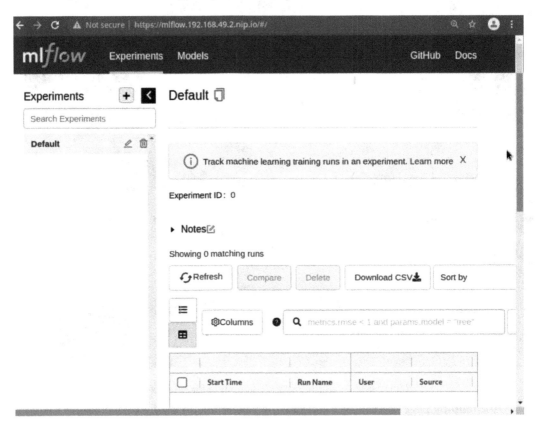

Figure 6.14 – MLflow experiment tracking page

The left-hand side of the **Experiments** screen contains the list of experiments, and the right-hand side displays the details of experiment runs. Think of the experiment as the data science project you are working on, such as fraud detection in consumer transactions, and the **Notes** section captures the combination of parameters, algorithms, and other information used to run the experiment.

6. Click on the **Models** tab to see the landing page of the model registry.

The **Models** tab contains the list of models in the registry, their versions, and their corresponding stages, which mention what environment the models are deployed in.

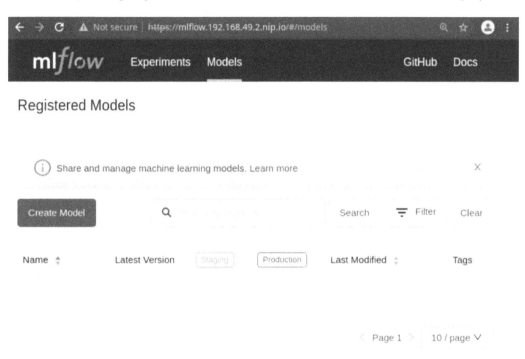

Figure 6.15 – MLflow model registry page

If you can open the MLflow URL and see the pages as described in the preceding steps, then you have just validated that MLflow is configured in your platform. The next step is to write a notebook that will train a basic model while recording the details in your MLflow server.

Using MLFlow as an experiment tracking system

In this section, you will see how the MLflow library allows you to record your experiments with the MLflow server. The custom notebook image, which you saw in the first part of this chapter, already has MLflow libraries packaged into the container. Please refer to the chapter6/requirements.txt file for the exact version of the MLflow library.

Before we start this activity, it is important to understand two main concepts: **experiment** and **run**.

An experiment is a logical name under which MLflow records and groups the metadata, for example, an experiment could be the name of your project. Let's say you are working on building a model for predicting credit card fraud for your retail customer. This could become the experiment name.

A run is a single execution of an experiment that is tracked in MLflow. A run belongs to an experiment. Each run may have a slightly different configuration, different hyperparameters, and sometimes, different datasets. You will tweak these parameters of the experiment in a Jupyter notebook. Each execution of model training is typically considered a run.

MLflow has two main methods of recording the experiment details. The first one, which is our preferred method, is to enable the auto-logging features of MLflow to work with your ML library. It has integration with Scikit, TensorFlow, PyTorch, XGBoost, and a few more. The second way is to record everything manually. You will see both options in the following steps.

These steps will show you how an experiment run or a model training can be recorded in MLflow while executing in a Jupyter notebook:

1. Log in to JupyterHub and make sure to select the custom container, for example, **Scikit v1.10 - Elyra Notebook Image**.

    Before you hit the **Start Server** button, add an environment variable by clicking on the **Add more variables** link. This variable may contain sensitive information such as passwords. MLflow needs this information to upload the artifacts to the Minio S3 server.

The landing page will look like the screenshot in *Figure 6.16*:

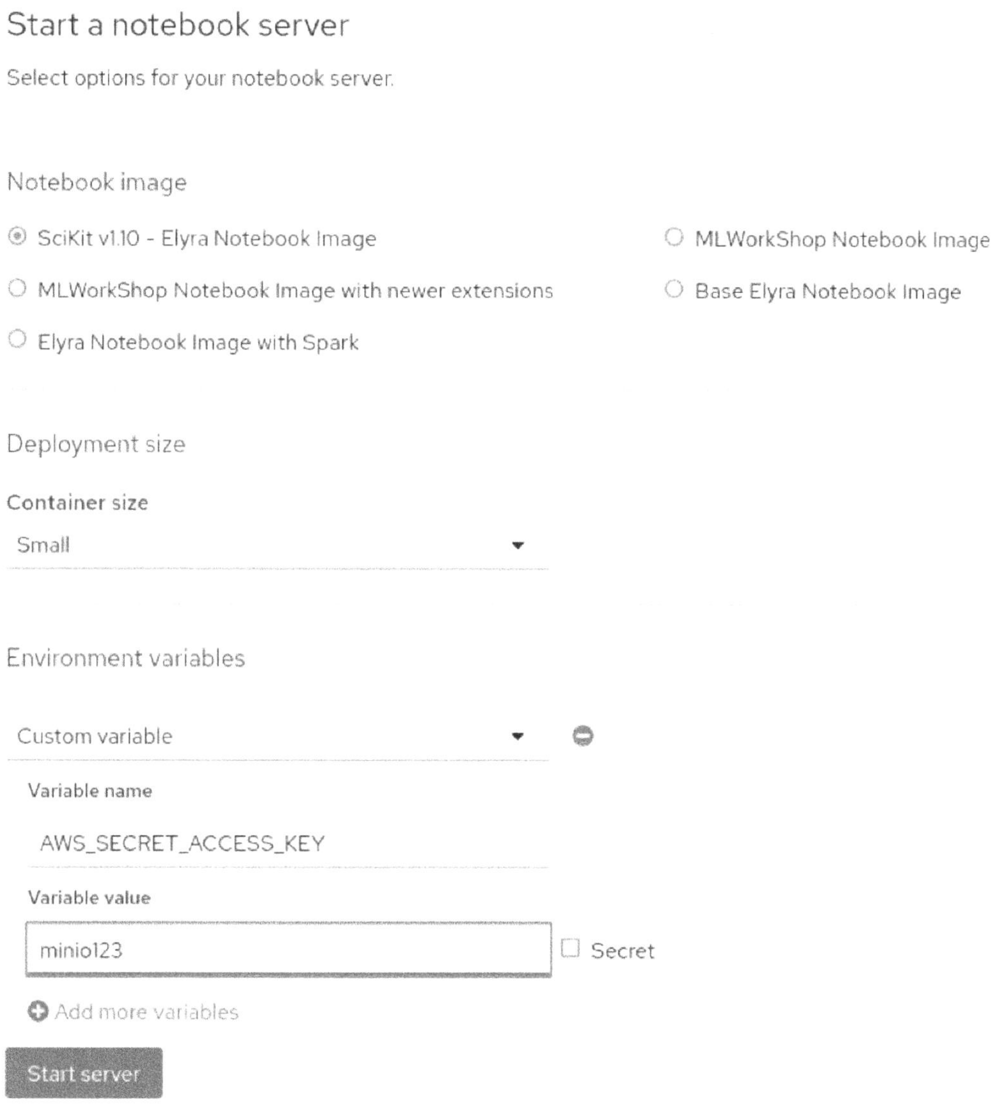

Start a notebook server

Select options for your notebook server.

Notebook image

◉ SciKit v1.10 - Elyra Notebook Image   ○ MLWorkShop Notebook Image

○ MLWorkShop Notebook Image with newer extensions   ○ Base Elyra Notebook Image

○ Elyra Notebook Image with Spark

Deployment size

Container size

Small   ▼

Environment variables

Custom variable   ▼   ⊖

Variable name

AWS_SECRET_ACCESS_KEY

Variable value

minio123   ☐ Secret

⊕ Add more variables

Start server

Figure 6.16 – JupyterHub with an environment variable

2. Open the notebook at `chapter6/hellomlflow.ipynb`. This notebook shows you how you can record your experiment data onto the MLflow server.

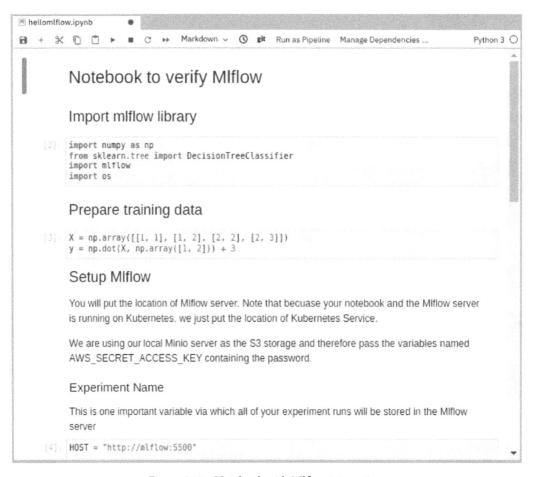

Figure 6.17 – Notebook with Mlflow integration

Note that at the first code cell, you have imported the MLflow library. In the second code cell, you have set up the location of the MLflow server through the `set_tracking_uri` method. Note that because your notebook and the MLflow server are running on Kubernetes, we just put the location of the Kubernetes Service that is stored in the `HOST` variable name and is being used in this method.

You then set the name of the experiment using the set_experiment method. This is one important variable through which all your experiment runs will be stored in the MLflow server.

The last method in this cell is sklearn.autolog, which is a way to tell MLflow that we are using the Scikit library for our training, and MLflow will record the data through Scikit APIs.

```
HOST = "http://mlflow:5500"

EXPERIMENT_NAME = "HelloMlFlow"

os.environ['MLFLOW_S3_ENDPOINT_URL']='http://minio-ml-workshop:9000'
os.environ['AWS_ACCESS_KEY_ID']='minio'
# os.environ['AWS_SECRET_ACCESS_KEY']='minio123'
os.environ['AWS_REGION']='us-east-1'
os.environ['AWS_BUCKET_NAME']='mlflow'

# Connect to local MLflow tracking server
mlflow.set_tracking_uri(HOST)

# Set the experiment name through which you will label all your exerpiments runs
mlflow.set_experiment(EXPERIMENT_NAME)

# enable autologging for scikit
mlflow.sklearn.autolog()
```

Figure 6.18 – Notebook cell with MLflow configuration

In the last cell of this notebook, you are using a simple DecisionTreeClassifier to train your model. Notice that this is quite a simple model and is used to highlight the capabilities of the MLflow server.

3. Run the notebook by selecting the **Run > Run all cells** menu option.

4. Log in to the MLflow server and click on the experiment name HelloMlFlow. The URL of MLflow will be like https://mlflow.192.168.61.72.nip.io with the IP address replaced as per your environment. As mentioned earlier in this chapter, you get this URL by listing the *ingress* objects of your Kubernetes cluster.

You will see the screen as shown in *Figure 6.19*:

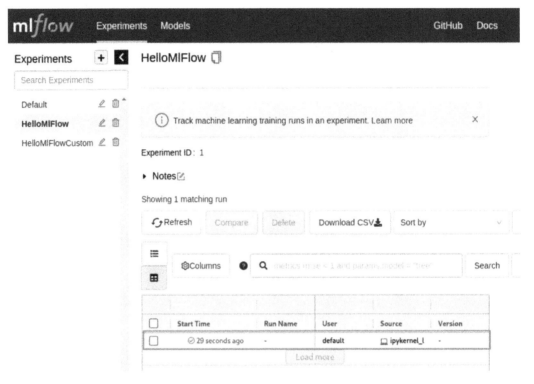

Figure 6.19 – MLflow experiments tracking screen showing an experiment run

You will notice that the table on the right-hand side contains one record. This is the experiment run you performed in *Step 6*. If you have executed your notebook multiple times with different parameters, each run will be recorded as a row in this table.

5. Click on the first row of the table.

    You will get to the details of the run you selected in the previous step. The screen will look like the screenshot in *Figure 6.20*:

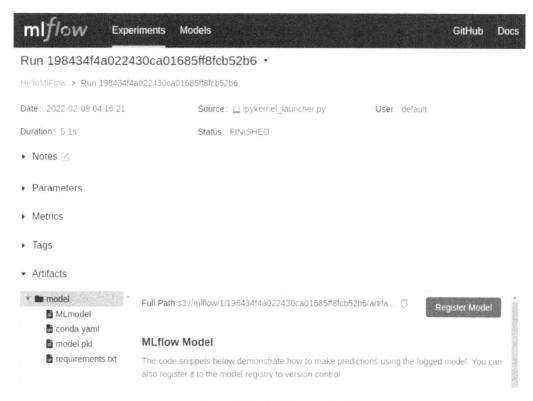

Figure 6.20 – MLflow run details

Let's understand the information that is available on this screen:

- **Parameters**: If you click on the little arrow next to **Parameters**, you will see that it has recorded the hyperparameters of your model training run. If you refer to the notebook code cell number 4, you will see that the parameters that we have used for `DecisionTreeClassifier` are recorded here too. One such example is the `max_depth` parameter, as shown in *Figure 6.21*:

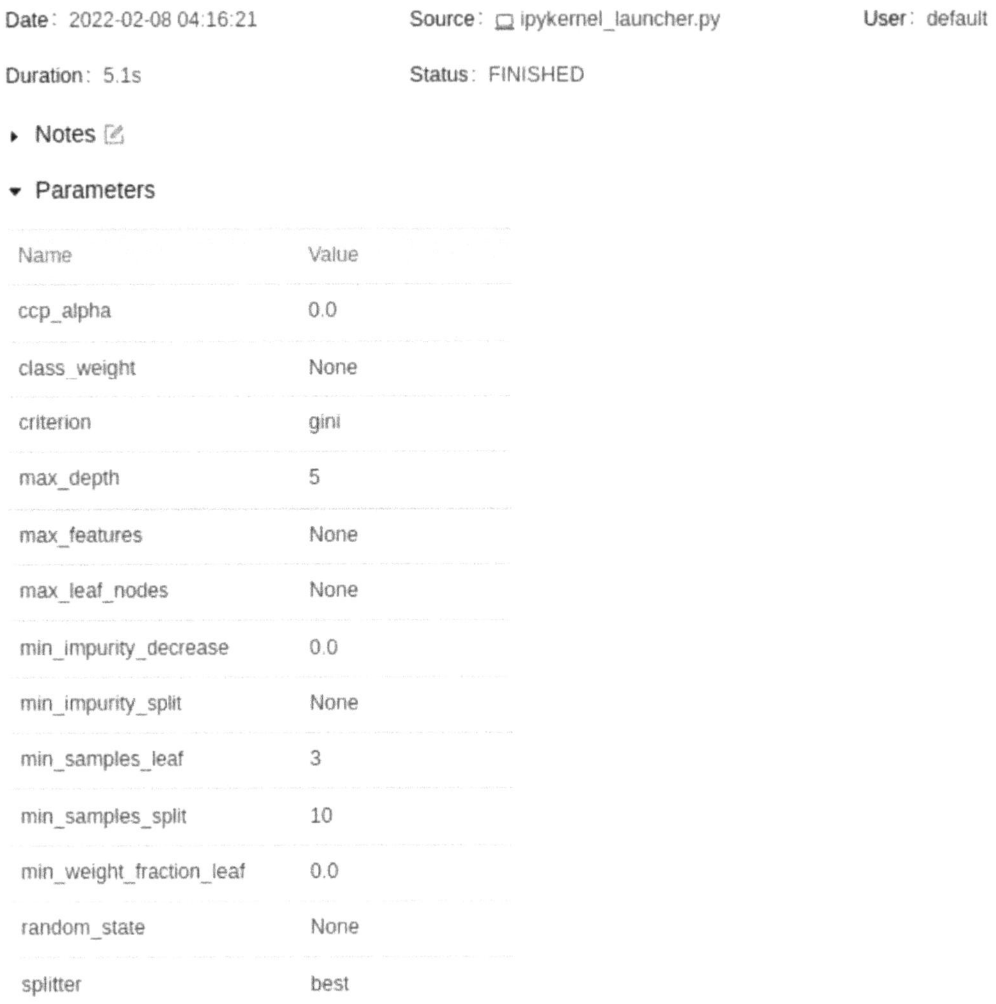

Date: 2022-02-08 04:16:21   Source: ⬜ ipykernel_launcher.py   User: default

Duration: 5.1s   Status: FINISHED

▸ Notes ✎

▾ Parameters

Name	Value
ccp_alpha	0.0
class_weight	None
criterion	gini
max_depth	5
max_features	None
max_leaf_nodes	None
min_impurity_decrease	0.0
min_impurity_split	None
min_samples_leaf	3
min_samples_split	10
min_weight_fraction_leaf	0.0
random_state	None
splitter	best

Figure 6.21 – MLflow run parameters

- **Metrics**: If you click on the little arrow next to **Metrics**, you will see that it has recorded the metrics for your model training run. You can see `training_accuracy` in the screenshot, as shown in *Figure 6.22*:

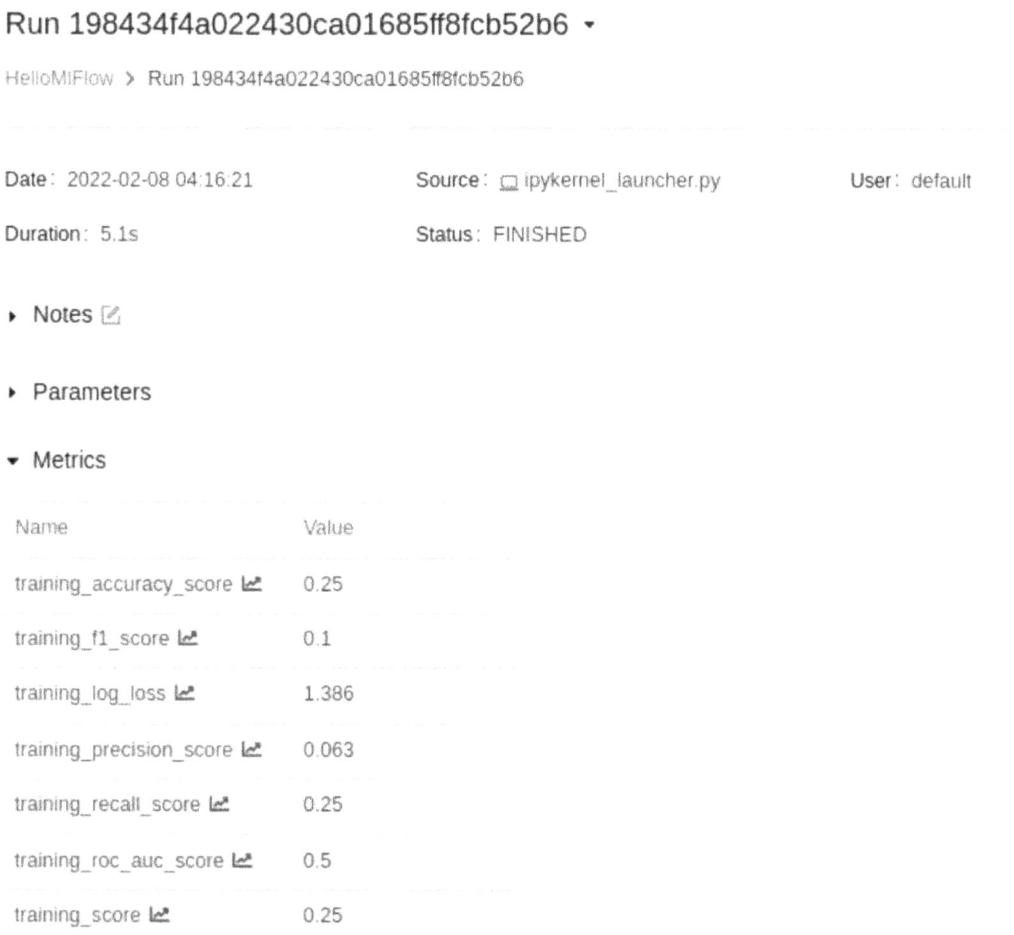

Run 198434f4a022430ca01685ff8fcb52b6 ▾

HelloMlFlow > Run 198434f4a022430ca01685ff8fcb52b6

Date : 2022-02-08 04:16:21   Source : ▢ ipykernel_launcher.py   User : default

Duration : 5.1s   Status : FINISHED

▸ Notes ✎

▸ Parameters

▾ Metrics

Name	Value
training_accuracy_score 📈	0.25
training_f1_score 📈	0.1
training_log_loss 📈	1.386
training_precision_score 📈	0.063
training_recall_score 📈	0.25
training_roc_auc_score 📈	0.5
training_score 📈	0.25

Figure 6.22 – MLflow run metrics

- **Tags**: If you click on the little arrow next to **Tags**, you will see the automatically associated tags (for example, `estimator_class`), which define the type of ML algorithm you have used. Note that you can add your own tags if needed. In the next section, we will show how to associate a custom tag for your run. *Figure 6.23* shows an example of tags:

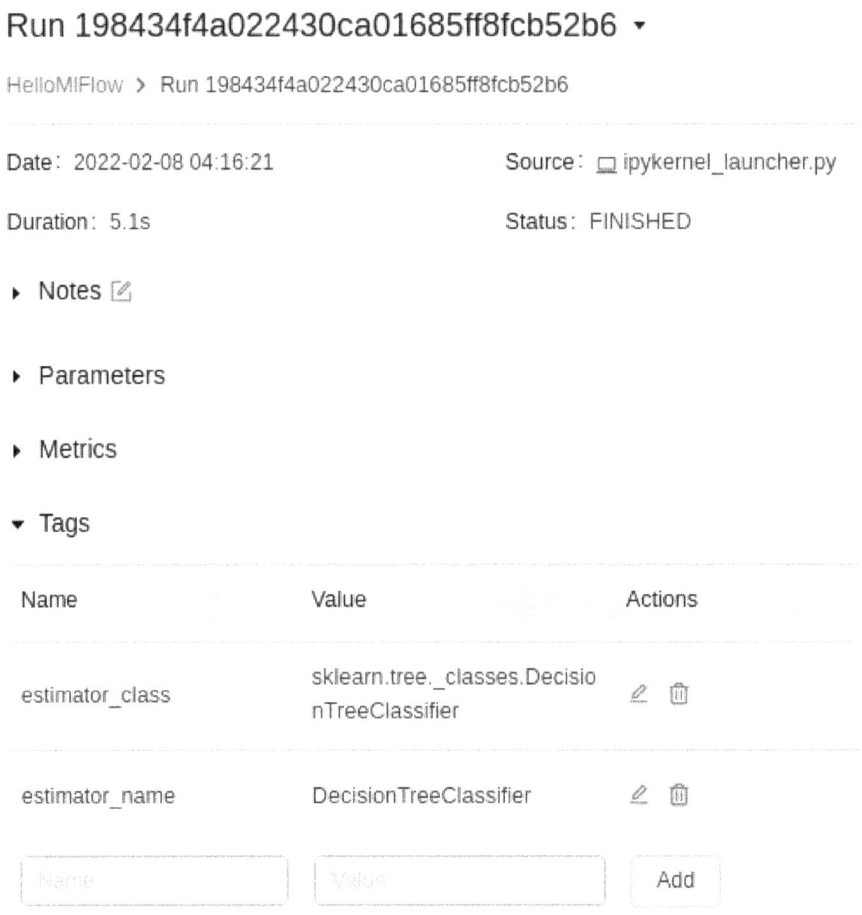

Run 198434f4a022430ca01685ff8fcb52b6 ▾

HelloMlFlow ❯ Run 198434f4a022430ca01685ff8fcb52b6

Date : 2022-02-08 04:16:21   Source : ▢ ipykernel_launcher.py

Duration : 5.1s   Status : FINISHED

▸ Notes ✎

▸ Parameters

▸ Metrics

▾ Tags

Name	Value	Actions
estimator_class	sklearn.tree._classes.DecisionTreeClassifier	✎ 🗑
estimator_name	DecisionTreeClassifier	✎ 🗑
Name	Value	Add

Figure 6.23 – MLflow run tags

- **Artifacts**: This section contains the artifacts associated with the run, such as the binary model file. Note that you can add your own artifacts here if needed. In the next section, we will show you how to associate an artifact with your run. Keep in mind that the artifacts are stored in the associated S3 bucket of your MLflow server. Note that the model binary is saved as a `model.pkl` file.

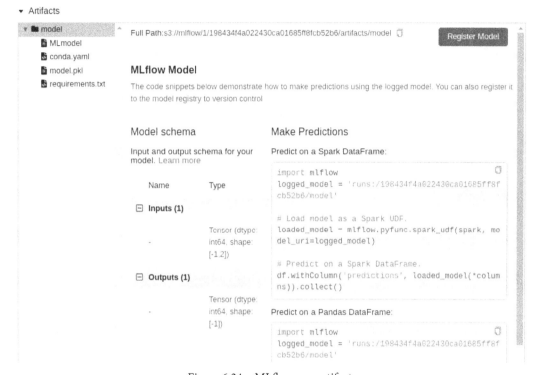

Figure 6.24 – MLflow run artifacts

6. To validate that these files are indeed stored in the S3 server, log in to the Minio server, select **Buckets**, and click on **Browse** button for the MLflow bucket. You will find a folder created with the name of your run. This name is displayed in the top-left corner of your experiment screen; consult the top-left corner of the preceding screen and you will see a label with a combination of 32 alphanumeric characters. This long number is your **run ID**, and you can see a folder label with a combination of 32 alphanumeric characters in your S3 bucket, as shown in the following screenshot. You can click on this link to find the artifacts stored on the S3 bucket:

Figure 6.25 – Minio bucket location

You have just successfully trained a model in JupyterHub and tracked the training run in MLflow.

You have seen how MLflow associates the data with each of your runs. You can even compare the data between multiple runs by selecting multiple runs from the table shown in *Step 6* and clicking on the **Compare** button.

Adding custom data to the experiment run

Now, let's see how we can add more data for each run. You will learn how to use the MLflow API to associate custom data with your experiment:

1. Start by firing up the Jupyter notebook as you did in the preceding section.

2. Open the notebook at `chapter6/hellomlflow-custom.ipynb`. This notebook shows you how you can customize the recording of your experiment data onto the MLflow server. The notebook is similar to the previous notebook, except for the code in cell number 6, which is shown in *Figure 6.26*. This code cell contains the functions that show how to associate data with your experiment:

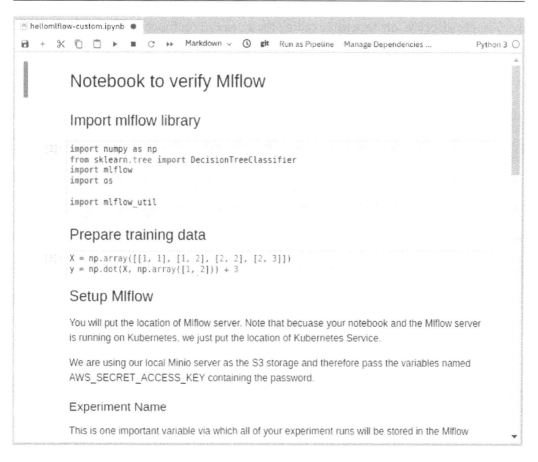

Figure 6.26 – MLflow customized data collection notebook

Let's understand these functions in the next few steps. The code snippet in code cell number 6 is as follows:

```
with mlflow.start_run(tags={     "mlflow.source.
git.commit" : mlflow_util.get_git_revision_hash()
,     "mlflow.source.git.branch": mlflow_util.get_git_
branch(),     "code.repoURL": mlflow_util.get_git_remote()
}) as run:        model.fit(X, y)     mlflow_util.record_
libraries(mlflow)    mlflow_util.log_metric(mlflow,
"custom_mteric", 1.0)    mlflow_util.log_param(mlflow,
"docker_image_name", os.environ["JUPYTER_IMAGE"])
```

The preceding code will include a custom tag labeled code.repoURL. This makes it easier to trace back the original source code that produced the model in a given experiment run.

3. You can associate any tags while calling the `start_run` function. Tag keys that start with *mlflow* are reserved for internal use. You can see that we have associated the GIT commit hash with the first property. This will help us in following through on what experiment belongs to what code version in your code repository.

You will find that the `code.repoURL` tag contains the Git repository location. You can add as many tags as you want. You can see the tags by going to the MLflow UI and opening the experiment. Note that the notebook has a different experiment name, and it is being referenced as `HelloMlFlowCustom`.

Note the **Git Commit** label at the top section of the page, and the custom tag name `code.repoURL` in the **Tags** section:

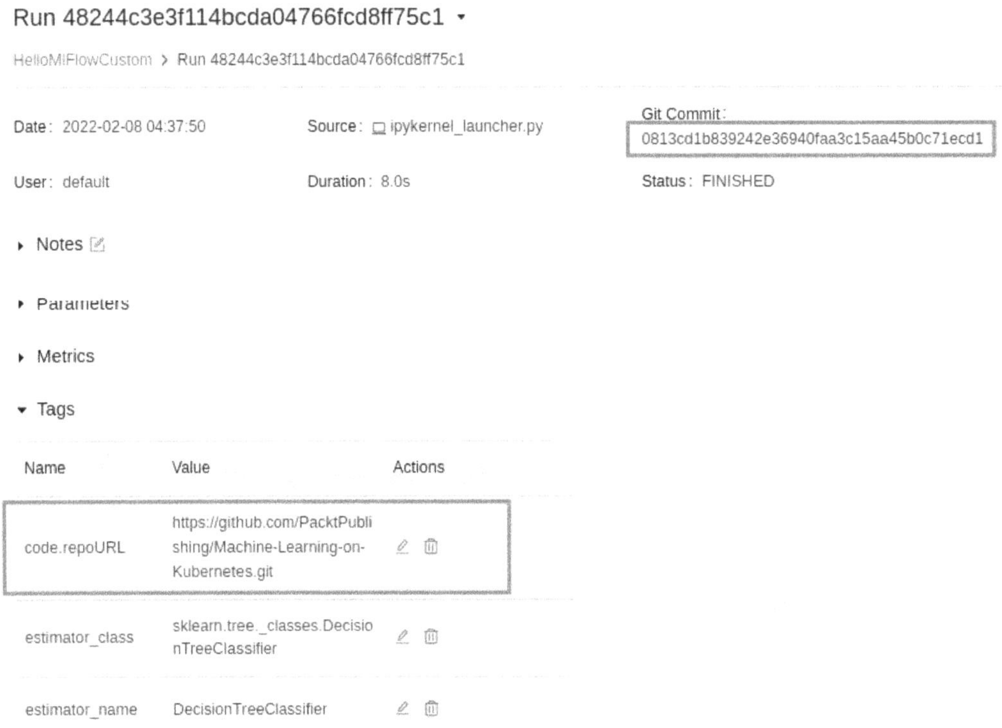

Run 48244c3e3f114bcda04766fcd8ff75c1 ▾

HelloMlFlowCustom > Run 48244c3e3f114bcda04766fcd8ff75c1

Date: 2022-02-08 04:37:50	Source: ▢ ipykernel_launcher.py	Git Commit: 0813cd1b839242e36940faa3c15aa45b0c71ecd1
User: default	Duration: 8.0s	Status: FINISHED

▸ Notes ☑

▸ Parameters

▸ Metrics

▾ Tags

Name	Value	Actions
code.repoURL	https://github.com/PacktPublishing/Machine-Learning-on-Kubernetes.git	✎ 🗑
estimator_class	sklearn.tree._classes.DecisionTreeClassifier	✎ 🗑
estimator_name	DecisionTreeClassifier	✎ 🗑

Figure 6.27 – MLflow custom tags

4. The second function that we have used is `record_libraries`. This is a wrapper function that internally uses the `mlflow.log_artifact` function to associate a file with the run. This utility function is capturing the `pip freeze` output, which gives the libraries in the current environment. The utility function then writes it to a file and uploads the file to the MLflow experiment. You can look at this, and all the other functions, in the `chapter6/mlflow_util.py` file.

You can see in the **Artifacts** section that a new file, pip_freeze.txt, is available, and it records the output of the pipe freeze command:

▾ Artifacts

▸ 📁 model
📄 pip_freeze.txt

Full Path:s3://mlflow/2/48244c3e3f114bcda04763fcd8ff75c1/artifacts/pip_fre... 📋
Size: 4.63KB

```
aiohttp==3.8.1
aiosignal==1.2.0
alembic==1.4.1
ansiwrap==0.8.4
anyio==3.4.0
argo-workflows==3.6.1
argon2-cffi==21.1.0
argon2-cffi-bindings==21.2.0
astroid==2.9.0
async-generator==1.10
async-timeout==4.0.1
attrdict==2.0.1
attrs==21.2.0
autopep8==1.5.5
Babel==2.9.1
backcall==0.2.0
backports.entry-points-selectable==1.1.1
backports.zoneinfo==0.2.1
beautifulsoup4==4.6.3
black==21.11b1
bleach==4.1.0
boto3==1.18.49
botocore==1.21.65
bump2version==1.0.1
bumpversion==0.6.0
cachetools==4.2.4
certifi==2021.10.8
certipy==0.1.3
cffi==1.15.0
charset-normalizer==2.0.8
click==8.0.3
```

Figure 6.28 – MLflow customized artifacts

5. The `log_metric` function records the metric name and its associated value. Note that the value for the metric is expected to be a number. For the sample code, you can see that we have just put a hardcoded value (`1`), however, in the real world, this would be a dynamic value that refers to something relative to each run of your experiment. You can find your custom metric in the **Metrics** section of the page:

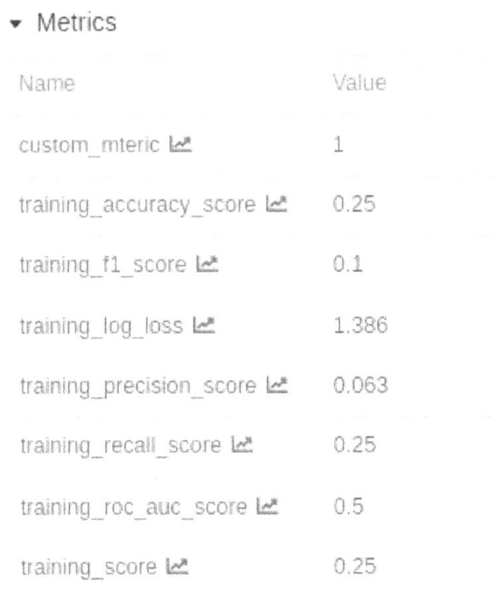

▾ Metrics

Name	Value
custom_mteric	1
training_accuracy_score	0.25
training_f1_score	0.1
training_log_loss	1.386
training_precision_score	0.063
training_recall_score	0.25
training_roc_auc_score	0.5
training_score	0.25

Figure 6.29 – MLflow customized metrics

6. The `log_param` function is like the `log_metric` function, but it can take any type of value against a given parameter name. For example, we have recorded the Docker image used by the Jupyter notebook. Recall that this is the custom image you built to be used by the data scientist team. You can see the following `docker_image_name` parameter that contains the desired value:

▾ Parameters

Name	Value
ccp_alpha	0.0
class_weight	None
criterion	gini
docker_image_name	quay.io/thoth-station/s2i-lab-elyra:v0.1.1
max_depth	5
max_features	None
max_leaf_nodes	None
min_impurity_decrease	0.0
min_impurity_split	None
min_samples_leaf	3
min_samples_split	10
min_weight_fraction_leaf	0.0
random_state	None
splitter	best

Figure 6.30 – MLflow customized parameters

You have used MLflow to track, add custom tags, and custom artifacts to an experiment run. In the next section, you will see the capabilities of MLflow as a model registry component. Let's dig in.

Using MLFlow as a model registry system

Recall that MLflow has a model registry feature. The registry provides the versioning capabilities for your models. Automation tools can get the models from the registry to deploy or even roll back your models across different environments. You will see in the later chapters that automation tools in our platform fetch the model from this registry via the API. For now, let's see how to use the registry:

1. Log in to the MLflow server by accessing the UI and clicking on the **Models** link. You should see the following screen. Click on the **Create Model** button:

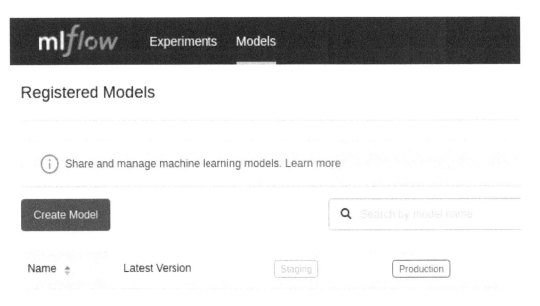

Figure 6.31 – MLflow registering a new model

2. Type a name for your model in the pop-up window, as shown in the following screenshot, and click on the **Create** button. This name could mention the name of the project that this model is serving:

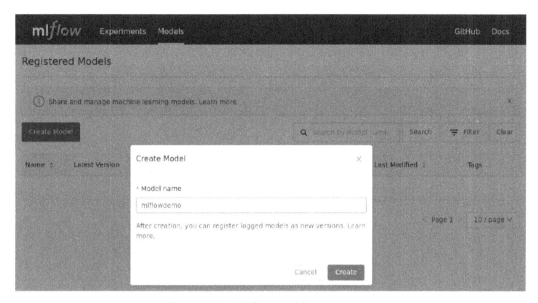

Figure 6.32 – MLflow model name prompt

3. Now, you need to attach a model file to this registered name. Recall from the preceding section that you have multiple *runs* in your *experiment*. Each run defines the set of configuration parameters and associated models with it. Select the experiment and run for which you want to register your model.

4. You will see a screen like the following. Select the **model** label in the **Artifacts** section, and you will notice a **Register Model** button on the right-hand side. Click on this button:

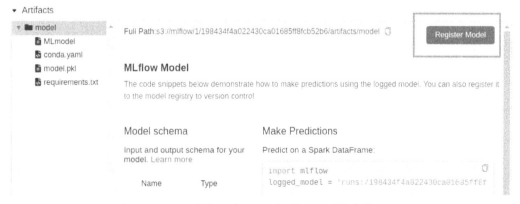

Figure 6.33 – MLflow showing the Register Model button

5. From the pop-up window, select the model name you created in *Step 1* and click on **Register**.

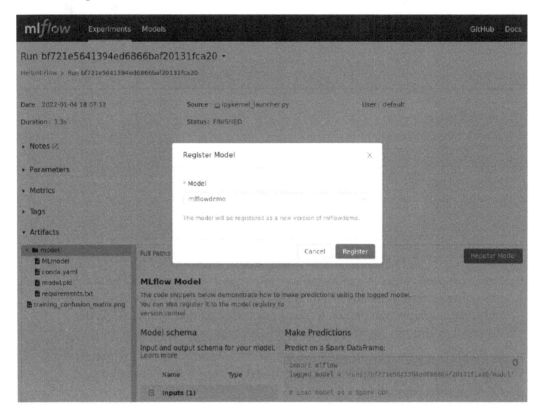

Figure 6.34 – Model name dialog when registering a model in MLflow

6. Go to the **Models** tab as mentioned in *Step 1* and you will see your model is registered in the MLflow registry. You will see the list as shown in the following screenshot. Click on the model name, for example, `mlflowdemo`:

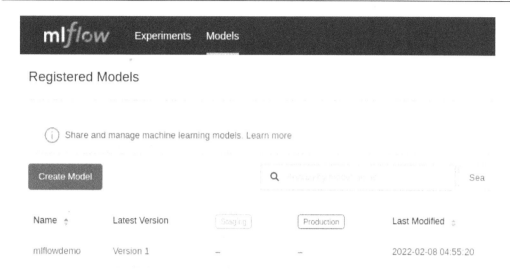

Figure 6.35 – MLflow showing the list of registered models and their versions

7. You will see the detail screen where you can attach the stage of the model as referred to by the **Stage** label. You can also edit other properties, and we will leave it to you to explore the data you can associate with this model:

Figure 6.36 – MLflow showing the buttons for promoting registered models to higher environments

Congratulations! You have just experienced using MLflow as a model registry! You have also seen how the model version can be promoted to the different stages of the lifecycle.

Summary

In this chapter, you have gained a better understanding of ML engineering and how it differs from data science. You have also learned about some of the responsibilities of ML engineers. You must take note that the definition of ML engineering and the role of ML engineers are still evolving, as more and more techniques are surfacing. One such technique that we will not talk about in this book is **online ML**.

You have also learned how to create a custom notebook image and use it to standardize notebook environments. You have trained a model in the Jupyter notebook while using MLflow to track and compare your model development parameters, training results, and metrics. You have also seen how MLflow can be used as a model registry and how to promote model versions to different stages of the lifecycle.

The next chapter will continue the ML engineering domain and you will package and deploy ML models to be consumed as an API. You will then automate the package and deploy the process using the tools available in the ML platform.

7
Model Deployment and Automation

In the previous chapter, you saw how the platform enables you to build and register the model in an autonomous fashion. In this chapter, we will extend the **machine learning (ML)** engineering domain to model deployment, monitoring, and automation of deployment activities.

You will learn how the platform provides the model packaging and deployment capabilities and how you can automate them. You will take the model from the registry, package it as a container, and deploy the model onto the platform to be consumed as an API. You will then automate all these steps using the workflow engine provided by the platform.

Once your model is deployed, it works well for the data it was trained upon. The real world, however, changes. You will see how the platform allows you to observe your model's performance. This chapter discusses the tools and techniques to monitor your model performance. The performance data could be used to decide whether the model needs retraining on the new dataset, or whether it is time to build a new model for the given problem.

In this chapter, you will learn about the following topics:

- Understanding model inferencing with Seldon Core

- Packaging, running, and monitoring a model using Seldon Core

- Understanding Apache Airflow

- Automating ML model deployments in Airflow

Technical requirements

This chapter includes some hands-on setup and exercises. You will need a running Kubernetes cluster configured with **Operator Lifecycle Manager**. Building such a Kubernetes environment is covered in *Chapter 3*, *Exploring Kubernetes*. Before attempting the technical exercises in this chapter, please make sure that you have a working Kubernetes cluster and **Open Data Hub** (**ODH**) is installed on your Kubernetes cluster. Installing ODH is covered in *Chapter 4*, *The Anatomy of a Machine Learning Platform*.

Understanding model inferencing with Seldon Core

In the previous chapter, you built the model. These models are built by data science teams to be used in production and serve the prediction requests. There are many ways to use a model in production, such as embedding the model with your customer-facing program, but the most common way is to expose the model as a REST API. The REST API can then be used by any application. In general, running and serving a model in production is called **model serving**.

However, once the model is in production, it needs to be monitored for performance and needs updating to meet the expected criteria. A hosted model solution enables you to not only serve the model but monitor its performance and generate alerts that can be used to trigger retraining of the model.

Seldon is a UK-based firm that created a set of tools to manage the model's life cycle. Seldon Core is an open source framework that helps expose ML models to be consumed as REST APIs. Seldon Core automatically exposes the monitoring statistics for the REST API, which can be consumed by **Prometheus**, the platform's monitoring component. To expose your model as a REST API in the platform, the following steps are required:

1. Write a language-specific wrapper for your model to expose as a service.

2. Containerize your model.

3. Define and deploy the model using the inference graph of your model using Seldon Deployment **custom resource (CR)** in Kubernetes

Next, we will see these three steps in detail.

Wrapping the model using Python

Let's see how you can apply the preceding steps. In *Chapter 6*, *Machine Learning Engineering*, you registered your experiment details and a model with the MLflow server. Recall that the model file was stored in the artifacts of MLflow and named `model.pkl`.

Let's take the model file and write a simple Python wrapper around it. The job of the wrapper is to use Seldon libraries to conveniently expose the model as a REST service. You can find the example of the wrapper in the code at `chapter7/model_deploy_ pipeline/model_build_push/Predictor.py`. The key component of this wrapper is a function named `predict` that will be invoked from an HTTP endpoint created by the Seldon framework. *Figure 7.1* shows a simple Python wrapper using a `joblib` model:

```python
import joblib


class Predictor(object):

    def __init__(self):
        self.model = joblib.load('model.pkl')

    def predict(self, data_array, column_names):
        return self.model.predict_proba(data_array)
```

Figure 7.1 – A Python language wrapper for the model prediction

The `predict` function receives a numpy array (`data_array`) and a set of column names (`column_names`), serialized from the HTTP request. The method returns the result of the prediction as either a numpy array or a list of values or bytes. There are many more methods available for the language wrapper and a full list is available at `https://docs.seldon. io/projects/seldon-core/en/v1.12.0/python/python_component. html#low-level-methods`. Note that in later chapters of this book, you will see a more thorough inferencing example that will have additional wrappers for data transformation before prediction. But, for this chapter, we keep it as simple as possible.

The language wrapper is ready, and the next stage is to containerize the model and the language wrapper.

Containerizing the model

What would you put in the container? Let's start with a list. You will need the model and the wrapper files. You will need the Seldon Python packages available in the container. Once you have all these packages, then you will use the Seldon services to expose the model. *Figure 7.2* shows a `Docker` file that is building one such container. This file is available in `Chapter 7/model_deployment_pipeline/model_build_push/Dockerfile.py`.

```
FROM registry.access.redhat.com/ubi8/python-38:1-77

WORKDIR /microservice
COPY base_requirements.txt /microservice/
RUN pip install -r base_requirements.txt

COPY requirements.txt /microservice/
RUN pip install -r requirements.txt

COPY Predictor.py   model.pkl /microservice/

CMD seldon-core-microservice $MODEL_NAME --service-type $SERVICE_TYPE --grpc-port ${GRPC_PORT} --metrics-port ${METRICS_PORT} --http-port ${HTTP_PORT}
```

Figure 7.2 – Docker file to package the model as a container

Now, let's understand the content of the Docker file:

- *Line 1* indicates the base container image for your model service. We have chosen the freely available image from Red Hat, but you can choose as per your convenience. This image could be your organization's base image with the standard version of Python and related software.

- In *Line 3*, we have created a `microservice` directory to place all the related artifacts in our container.

- In *Line 4*, the first file we need to build the container is `base_requirements.txt`. This file contains the packages and dependencies for the Seldon Core system. You can find this file at `chapter7/model_deployment_pipeline/model_build_push/base_requirements.txt`. In this file, you will see that Seldon Core packages and `joblib` packages have been added.

Figure 7.3 shows the `base_requirements.txt` file:

Figure 7.3 – File adding Seldon and Joblib to the container

- *Line 5* is using the `base_requirements.txt` file to install the Python packages onto the container.

- In *Lines 7* and *8*, when you are training the model, you may use different packages. During inferencing, some of the packages may be needed; for example, if you have done input data scaling before model training using a library, you may need the same library to apply the scaling at inference time.

In *Chapter 6, Machine Learning Engineering*, you registered your experiment details and a model with the MLflow server. Recall that the model file was stored in the artifacts along with a file containing packages used to train the model named `requirements.txt`. Using the `requirements.txt` file generated by MLflow, you can install the packages required to run your model, or you may choose to add these dependencies on your own to a custom file. *Figure 7.4* shows the MLflow snapshot referred to in *Chapter 6, Machine Learning Engineering*. You can see the `requirements.txt` file here next to the `model.pkl` file.

Run a4d19698f1094a9eb6b66d33df943363 ▾

HelloMlFlow > Run a4d19698f1094a9eb6b66d33df943363

Date : 2022-05-15 17:50:03   Source : ▢ ipykernel_launcher.py   User : default

Duration : 2.7s   Status : FINISHED

▸ Notes ✎

▸ Parameters

▸ Metrics

▸ Tags

▾ Artifacts

▾ ■ model   Full Path:s3://mlflow/2/a4d19698f1094a9eb6b66d33df943363/artifacts/model ▢   Register Model
   ■ MLmodel
   ■ conda.yaml   **MLflow Model**
   ■ model.pkl
   ■ requirements.txt   The code snippets below demonstrate how to make predictions using the logged model. You can also register it to the
 ■ training_confusion_matrix.png   model registry to version control

Figure 7.4 – MLflow run artifacts

Line 10: You add the language wrapper files and the model files to the container.

Line 11: Here, you are using the `seldon-core-microservice` server to start the inferencing server. Notice that the parameters have been passed here, and in the next section, you will see how we can pass these parameters:

- **MODEL_NAME**: This is the name of the Python class in the language wrapper containing the model.

- **SERVICE_TYPE**: This parameter contains the type of service being created here in the inference pipeline. Recall that an inference pipeline may contain the model execution or data transformation or it may be an outlier-detector. For model execution, the value of this parameter will be `MODEL`.

- **GRPC_PORT**: The port at which the **Google Remote Procedure Call (gRPC)** endpoint will listen for model inference.

- **METRICS_PORT**: The port at which the service performance data will be exposed. Note that this is the performance data for the service and not the model.

- **HTTP_NAME**: The HTTP port where will you serve the model over HTTP.

Now, we have a container specification in the form of the Docker file. Next, we will see how to deploy the container on the Kubernetes platform using the Seldon controller.

Deploying the model using the Seldon controller

Our ML platform provides a Seldon controller, a piece of software that runs as a pod and assists in deploying the containers you built in the preceding section. Note that the controller in our platform is the extension of the existing Seldon operator. At the time of writing, the Seldon operator was not compatible with Kubernetes version 1.22, so we have extended the existing operator to work with the latest and future versions of the Kubernetes platform.

Refer to *Chapter 4*, *The Anatomy of a Machine Learning Platform*, where you learned about installing ODH and how it works on the Kubernetes cluster. In an equivalent manner, the Seldon controller is also installed by the ODH operator. The `manifests/ml-platform.yaml` file has the configuration for installing the Seldon controller. *Figure 7.5* shows the settings:

```
  - kustomizeConfig:
      repoRef:
        name: manifests
        path: manifests/odhseldon/cluster
    name: odhseldon
```

Figure 7.5 – MLFlow section of the manifest file

Let's verify whether the Seldon controller is running correctly in the cluster:

```
kubectl get pods -n ml-workshop | grep -i seldon
```

You should see the following response:

```
$kubectl get pods -n ml-workshop | grep seldon
seldon-controller-manager-7f67f4985b-vxvjt                    1/1      Running
```

Figure 7.6 – Seldon controller pod

The Seldon controller pod is installed by the ODH operators, which watch for the Seldon Deployment CR. This schema for this resource is defined by the Seldon Deployment **custom resource definition (CRD)**; you can find the CRD at `manifests/odhseldon/cluster/base/seldon-operator-crd-seldondeployments.yaml`. Once you create the Seldon Deployment CR, the controller deploys the pods associated with the CR. *Figure 7.7* shows this relationship:

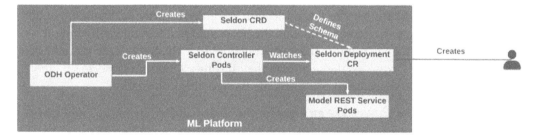

Figure 7.7 – Components of the platform for deploying Seldon services

Let's see the different components of the Seldon Deployment CR. You can find one simple example in `chapter7/manual_model_deployment/SeldonDeploy.yaml`.

The Seldon Deployment CR contains all the information that is required by the Seldon controller to deploy your model on the Kubernetes cluster. There are three main sections in the Seldon Deployment CR:

- **General information**: This is the section that describes `apiVersion`, `kind`, and other Kubernetes-related information. You will define the labels and name of the Seldon Deployment as any other Kubernetes object. You can see in the following screenshot that it contains the labels and annotations for the object:

```
 1 ▶    apiVersion: machinelearning.seldon.io/v1
 2      kind: SeldonDeployment
 3    ⊟metadata:
 4    ⊟  labels:
 5    ⊟    app: seldon
 6    ⊟  name: model-test
 7    ⊟spec:|
 8    ⊟  annotations:
 9    |    project_name: test
10    ⊟    deployment_version: v1
11         name: model-test-spec
```

Figure 7.8 – Seldon Deployment – Kubernetes-related information

- **Container specifications**: The second section is where you provide details about the container location, the deployment, and the horizontal pod scaling configuration of your service. Note that this is the same container that you built in the preceding section. *Figure 7.7* contains the section of the `chapter7/manual_model_deployment/SeldonDeploy.yaml` file that has this information.

Notice that `containers` take an array for the `image` object, so you can add more images to it. The `image` key will have the location of your container. The `env` array defines the environment variables that will be available for the pod. Recall that, in our Docker file in the previous section, these variables have been used. `MODEL_NAME` has a value of `Predictor`, which is the name of the class you have used as a wrapper. `SERVICE_TYPE` has a value of `MODEL`, which mentions the type of service this container provides.

The last part has hpaSpec, which the Seldon controller will translate onto the **Kubernetes Horizontal Pod Autoscaler** object. Through these settings, you can control the scalability of your pods while serving inferencing calls. For the following example, maxReplicas is set to 1, so there will not be any new pods, but you can control this value for each deployment. The scalability will kick in if the CPU utilization goes beyond 80% for the pods in the following example; however, because maxReplica is 1, there will not be any new pods created.

```
12    predictors:
13      - componentSpecs:
14        - spec:
15          containers:
16            - image: <INSERT CONTAINER LOCATION HERE>
17              imagePullPolicy: Always
18              name: model-test-predictor
19              env:
20                - name: MODEL_NAME
21                  value: "Predictor"
22                - name: SERVICE_TYPE
23                  value: MODEL
24                - name: GRPC_PORT
25                  value: "5005"
26                - name: METRICS_PORT
27                  value: "6005"
28                - name: HTTP_PORT
29                  value: "9000"
30          hpaSpec:
31            maxReplicas: 1
32            metrics:
33              - resource:
34                  name: cpu
35                  targetAverageUtilization: 80
36                type: Resource
37            minReplicas: 1
```

Figure 7.9 – Seldon Deployment – Seldon service containers

- **Inference graph**: The section under the graph key builds the inference graph for your service. An inference graph will have different nodes and you will define what container will be used at each node. You will see there is a children key that takes an array of objects through which you define your inference graph. For this example, graph has only one node and the children key has no information associated with it; however, in the later chapters, you will see how to build the inference graph with more nodes.

The remaining fields under the graph define the first node of your inference graph. The name field has the value that corresponds to the name you have given in the containers section. Note that this is the key through which Seldon knows what container would be serving at this node of your inference graph.

The other important part is the logger section. Seldon can automatically forward the request and response to the URL mentioned under the logger section. The capability of forwarding the request and response can be used for a variety of scenarios, such as storing the payload for audit/legal reasons or applying data drift algorithms to trigger retraining or anything else. Note that Seldon can also forward to Kafka if needed, but this is outside the scope of this book.

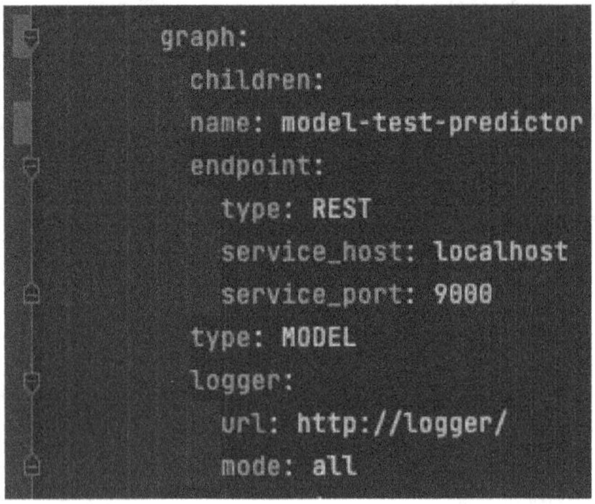

```
graph:
  children:
  name: model-test-predictor
  endpoint:
    type: REST
    service_host: localhost
    service_port: 9000
  type: MODEL
  logger:
    url: http://logger/
    mode: all
```

Figure 7.10 – Seldon Deployment – inference graph

Once you create the Seldon Deployment CR using the routine kubectl command, the Seldon controller will deploy the pods, and the model will be available for consumption as a service.

Next, we'll move on to packaging and deploying the basic model that you built in *Chapter 6, Machine Learning Engineering*.

Packaging, running, and monitoring a model using Seldon Core

In this section, you will package and build the container from the model file you built in *Chapter 6*, *Machine Learning Engineering*. You will then use the Seldon Deployment to deploy and access the model. Later in this book, you will automate the process, but to do it manually, as you'll do in this section, we will further strengthen your understanding of the components and how they work.

Before you start this exercise, please make sure that you have created an account with a public Docker registry. We will use the free `quay.io` as our registry, but you are free to use your preferred one:

1. Let's first verify that MLflow and Minio (our S3 server) are running in our cluster:

   ```
   kubectl get pods -n ml-workshop | grep -iE 'mlflow|minio'
   ```

   You should see the following response:

   ```
   minio-ml-workshop-6b84fdc7c4-z7hxk        1/1    Running
   mlflow-7b954c468f-7q2ws                   2/2    Running
   mlflow-db-0                               1/1    Running
   ```

   Figure 7.11 – MLflow and Minio are running on the platform

2. Get the ingress list for MLflow, and log in to MLflow using the `mlflow` URL available from the following output:

   ```
   kubectl get ingresses.networking.k8s.io -n ml-workshop
   ```

   You should see the following response:

   ```
   NAME                   CLASS    HOSTS
   ap-airflow2            nginx    airflow.192.168.61.72.nip.io
   grafana                nginx    grafana.192.168.61.72.nip.io
   jupyterhub             nginx    jupyterhub.192.168.61.72.nip.io
   minio-ml-workshop-ui   nginx    minio.192.168.61.72.nip.io
   mlflow                 nginx    mlflow.192.168.61.72.nip.io
   ```

   Figure 7.12 – ingress in your Kubernetes cluster

3. Once you are in the MLflow UI, navigate to the experiment that you recorded in *Chapter 6, Machine Learning Engineering*. The name of the experiment is **HelloMIFlow**.

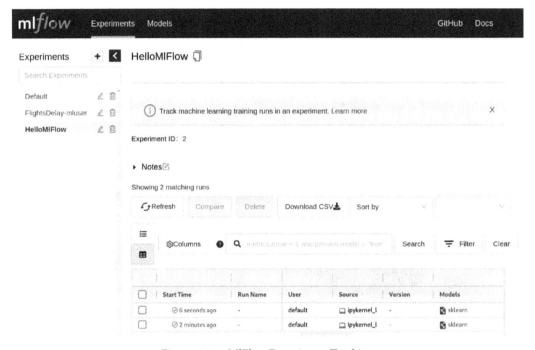

Figure 7.13 – MlFlow Experiment Tracking

4. Select the first run from the right-hand panel to get to the detail page of the run. From the **Artifacts** section, click on `model.pkl` and you will see a little download arrow icon to the right. Use the icon to download the **model.pkl** and `requirements.txt` files from this screen.

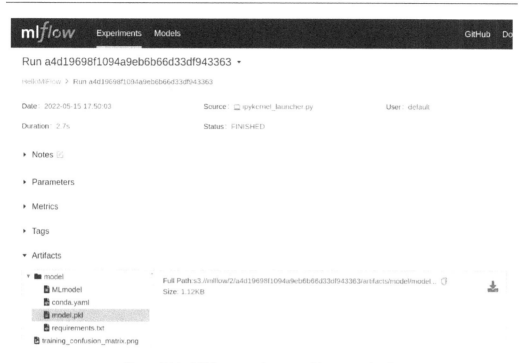

Figure 7.14 – MLflow experiment tracking – run details

5. Go to the folder where you have cloned the code repository that comes with this book. If you have not done so, please clone the `https://github.com/PacktPublishing/Machine-Learning-on-Kubernetes.git` repository on your local machine.

6. Then, go to the `chapter7/model_deploy_pipeline/model_build_push` folder and copy the two files downloaded in the previous step to this folder. In the end, this folder will have the following files:

```
build_push_image.py
Predictor.py
Dockerfile
base_requirements.txt
model.pkl
requirements.txt
```

Figure 7.15 – Sample files to package the model as a container

> **Note**
> The last two files are the ones that you have just copied. All other files are coming from the code repository that you have cloned.

Curious people will note that the `requirements.txt` file that you have downloaded from the MLFlow server contains the packages required while you run the notebook for model training. Not all of these packages (`mlflow`, for example) will be needed to execute the saved model. To keep things simple, we will add all of them to our container.

7. Now, let's build the container on the local machine:

```
docker build -t hellomlflow-manual:1.0.0 .
```

You should see the following response:

```
$docker build -t hellomlflow-manual:1.0.0 .
[+] Building 38.5s (12/12) FINISHED
 => [internal] load build definition from Dockerfile
 => => transferring dockerfile: 573B
 => [internal] load .dockerignore
 => => transferring context: 2B
 => [internal] load metadata for registry.access.redhat.com/ubi8/python-38:1-77
 => [1/7] FROM registry.access.redhat.com/ubi8/python-38:1-77@sha256:67bda161ea329167fee43a8687d26d3cd7dc717b4272b8ed244628fb4f4864e8
 => => resolve registry.access.redhat.com/ubi8/python-38:1-77@sha256:67bda161ea329167fee43a8687d26d3cd7dc717b4272b8ed244628fb4f4864e8
 => => sha256:2b4bac4b01df98158b7ead4352a24a79184ea8c8a8108294cb437ee95a5a34d5 1.37kB / 1.37kB
 => => sha256:5fa3b1ecd720b5c0d4dc1de64c48562ee87627446ad6920c3b316786c26c3730f 8.48kB / 8.48kB
 => => sha256:67bda161ea329167fee43a8687d26d3cd7dc717b4272b8ed244628fb4f4864e8 1.48kB / 1.48kB
 => [internal] load build context
 => => transferring context: 1.88kB
 => [2/7] WORKDIR /microservice
 => [3/7] COPY base_requirements.txt /microservice/
 => [4/7] RUN pip install -r base_requirements.txt
 => [5/7] COPY requirements.txt /microservice/
 => [6/7] RUN pip install -r requirements.txt
 => [7/7] COPY Predictor.py  model.pkl /microservice/
 => exporting to image
 => => exporting layers
 => => writing image sha256:76f3039c8a91bb8e34daf5ece75f33fdb67dda50c4845892f7789c598cd94254
 => => naming to docker.io/library/hellomlflow-manual:1.0.0
```

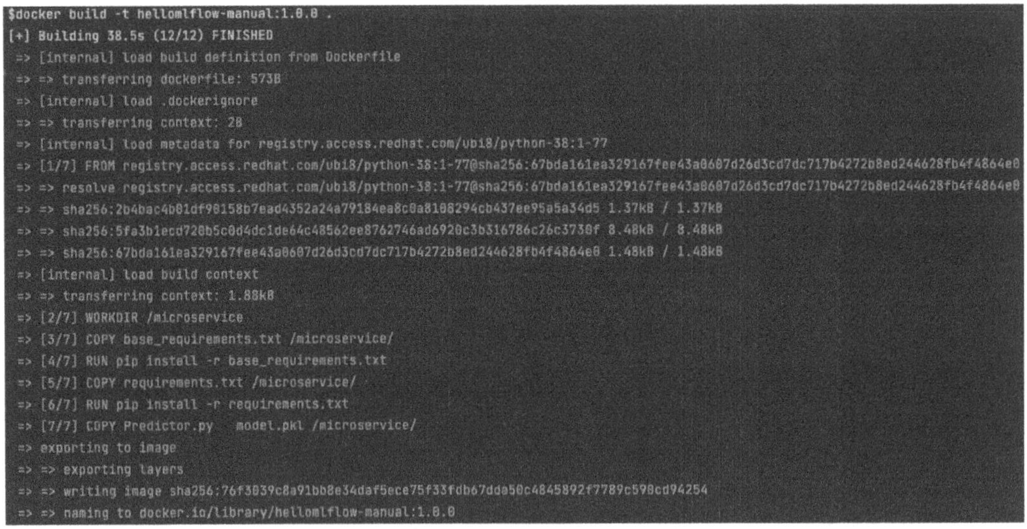

Figure 7.16 – Packaging the model as a container

8. The next step is to tag the container and push it to the repository of your choice. Before you push your image to a repository, you will need to have an account with an image registry. If you do not have one, you can create one at `https://hub.docker.com` or `https://quay.io`. Once you have created your registry, you can run the following commands to tag and push the image:

```
docker tag hellomlflow-manual:1.0.0 <DOCKER_REGISTRY>/
hellomlflow-manual:1.0.0
```

```
docker push <DOCKER_REGISTRY> /hellomlflow-manual:1.0.0
```

You should see the following response. You will notice that, in the following screenshot, we refer to `quay.io/ml-on-k8s` as our registry:

```
$docker push quay.io/ml-on-k8s/hellomlflow-manual:1.0.0
The push refers to repository [quay.io/ml-on-k8s/hellomlflow-manual]
447350ea4072: Pushed
1a4cacc07eb4: Pushed
ba0019486e6f: Pushed
26e9f063ed76: Pushed
cde643b1a2b1: Pushed
33c2bc928dd5: Pushed
bf7c04d1de2d: Mounted from ml-on-k8s/container-model
7b17276847a2: Mounted from ml-on-k8s/container-model
558b534f4e1b: Mounted from ml-on-k8s/container-model
3ba8c926eef9: Mounted from ml-on-k8s/container-model
352ba846236b: Mounted from ml-on-k8s/container-model
1.0.0: digest: sha256:ec6b6c5f8a38aea72a65df5d3bf9d1bef144dca0bba9efb3c7b2dd3cacce83a1 size: 2627
```

Figure 7.17 – Pushing the model to a public repository

9. Now that your container is available in a registry, you will need to use the Seldon Deployment CR to deploy it as a service. Open the `chapter7/manual_model_deployment/SeldonDeploy.yaml` file and adjust the location of the image.

   You can see the file after I have modified *line 16* (as per my image location) as follows:

```
manual_model_deployment/SeldonDeploy.yaml
3  metadata:
4    labels:
5      app: seldon
6    name: model-test
7  spec:
8    annotations:
9      project_name: test
10     deployment_version: v1
11   name: model-test-spec
12   predictors:
13     - componentSpecs:
14       - spec:
15           containers:
16             - image: quay.io/ml-on-k8s/hellomlflow-manual:1.0.0
17               imagePullPolicy: Always
18               name: model-test-predictor
19               env:
20                 - name: MODEL_NAME
21                   value: "Predictor"
22                 - name: SERVICE_TYPE
23                   value: MODEL
24                 - name: GRPC_PORT
```

Figure 7.18 – Seldon Deployment CR with the image location

10. Let's deploy the model as a service by deploying the `chapter7/manual_model_ deployment/SeldonDeploy.yaml` file. Run the following command:

```
kubectl create -f chapter7/manual_model_deployment/
SeldonDeploy.yaml -n ml-workshop
```

You should see the following response:

```
$kubectl create -f chapter7/manual_model_deployment/SeldonDeploy.yaml -n ml-workshop
seldondeployment.machinelearning.seldon.io/model-test created
```

Figure 7.19 – Creating the Seldon Deployment CR

11. Validate that the container is in a running state. Run the following command:

```
kubectl get pod -n ml-workshop | grep model-test-
predictor
```

You will note that the name that you have put in the `graph` section of the `SeldonDeploy.yaml` file (`model-test-predictor`) is part of the container name.

You should see the following response:

```
$kubectl get pod -n ml-workshop | grep model-test-predictor
model-test-predictor-0-model-test-predictor-76d5c5b8f-d9nqf       2/2       Running
```

Figure 7.20 – Validating the pod after the Seldon Deployment CR

12. Great! You have a model running as a service. Now, let's see what is in the pod created for us by the Seldon controller. Run the following command to get a list of containers inside our pod:

```
export POD_NAME=$(kubectl get pod -o=custom-
columns=NAME:.metadata.name -n ml-workshop | grep model-
test-predictor)
```

```
kubectl get pods $POD_NAME -o jsonpath='{.spec.
containers[*].name}' -n ml-workshop
```

You should see the following response:

```
$export POD_NAME=$(kubectl get pod -o=custom-columns=NAME:.metadata.name -n ml-workshop | grep model-test-predictor)
$kubectl get pods $POD_NAME -o jsonpath='{.spec.containers[*].name}' -n ml-workshop
model-test-predictor seldon-container-engine$
```

Figure 7.21 – Containers inside the Seldon pod

You will see that there are two containers. One is `model-test-predictor`, which is the image that we have built, and the second container is `seldon-container-engine`, which is the Seldon server.

The `model-test-predictor` container has the model and is using the language wrapper to expose the model over HTTP and gRPC. You can use the following command to see the logs and what ports have been exposed from `model-test-predictor`:

```
kubectl logs -f $POD_NAME -n ml-workshop -c model-test-predictor
```

You should see the following response (among other logs):

```
- seldon_core.microservice:main:374 - INFO:    Importing Predictor
- seldon_core.microservice:main:463 - INFO:    REST gunicorn microservice running on port 9000
- seldon_core.microservice:main:557 - INFO:    REST metrics microservice running on port 6005
- seldon_core.microservice:main:567 - INFO:    Starting servers
```

Figure 7.22 – Containers log showing the ports

You can see that the servers are ready to take the calls on `9000` for HTTP and on `6005` for the metrics server. This metrics server will have the Prometheus-based monitoring data exposed on the `/prometheus` endpoint. You can see this in the following portion of the log:

```
seldon_core.microservice:load_annotations:163 - INFO:    Found annotation prometheus.io/path:/prometheus
seldon_core.microservice:load_annotations:163 - INFO:    Found annotation prometheus.io/scrape:true
seldon_core.microservice:load_annotations:163 - INFO:    Found annotation seldon.io/svc-name:model-test
```

Figure 7.23 – Containers log showing the Prometheus endpoint

The second container is `seldon-container-engine`, which does the orchestration for the inference graph and forwards the payloads to the service configured by you in the `logger` section of the Seldon Deployment CR.

13. In this step, you will find out what Kubernetes objects your Seldon Deployment CR has created for you. A simple way to find out is by running the command as follows. This command depends on the Seldon controller labeling the objects it creates with the label key as `seldon-deployment-id`, and the value is the name of your Seldon Deployment CR, which is `model-test`:

```
kubectl get all  -l seldon-deployment-id=model-test -n ml-workshop
```

You should see the following response:

```
$kubectl get all   -l seldon-deployment-id=model-test -n ml-workshop
NAME                                                     READY   STATUS    RESTARTS   AGE
pod/model-test-predictor-0-model-test-predictor-76d5c5b8f-d9nqf   2/2     Running   0          5h21m

NAME                                             TYPE        CLUSTER-IP     EXTERNAL-IP   PORT(S)           AGE
service/model-test                               ClusterIP   10.98.126.3    <none>        8000/TCP,5001/TCP  5h21m
service/model-test-predictor-model-test-predictor ClusterIP   10.102.2.177   <none>        9000/TCP,9500/TCP  5h21m

NAME                                                     READY   UP-TO-DATE   AVAILABLE   AGE
deployment.apps/model-test-predictor-0-model-test-predictor   1/1     1            1           5h21m

NAME                                                           DESIRED   CURRENT   READY   AGE
replicaset.apps/model-test-predictor-0-model-test-predictor-76d5c5b8f   1         1         1       5h21m

NAME                                                                      REFERENCE
horizontalpodautoscaler.autoscaling/model-test-predictor-0-model-test-predictor   Deployment/model-test-predictor-0-model-test-predictor
```

Figure 7.24 – Kubernetes objects created by the Seldon controller

You can see that there are Deployment objects, services, and **Horizontal Pod Autoscalers** (**HPA**) objects created for you for the Seldon controller using the configuration that you have provided in the Seldon Deployment CR. The deployment ends up creating pods and a replica set for your pods. The Seldon controller made it easy to deploy our model on the Kubernetes platform.

14. You may have noticed that there is no ingress object created by the Seldon Deployment CR. Let's create the ingress object so that we can call our model from outside the cluster by running the command as follows. The ingress object is created by the file in `chapter7/manual_model_deployment/Ingress.yaml`. Make sure to adjust the `host` value as per your configuration, as you have done in earlier chapters. You will also notice that the ingress is forwarding traffic to port `8000`. Seldon provides the listener to this port, which orchestrates the inference call. This service is available in the container named `seldon-container-engine`:

```
kubectl create -f chapter7/manual_model_deployment/
Ingress.yaml -n ml-workshop
```

You should see the following response:

```
$kubectl create -f chapter7/manual_model_deployment/Ingress.yaml -n ml-workshop
ingress.networking.k8s.io/model-test created
```

Figure 7.25 – Creating ingress objects for our service

Validate that the ingress has been created by issuing the following command:

```
kubectl get ingress -n ml-workshop | grep model-test
```

You should see the following response:

```
$kubectl get ingress -n ml-workshop | grep model-test
model-test              nginx    model-test.192.168.61.72.nip.io    localhost    80
```

Figure 7.26 – Validating the ingress for our service

15. Since our Seldon Deployment CR has referenced a logger URL, you will deploy a simple HTTP echo server that will just print the calls it received. This will assist us in validating whether the payloads have been forwarded to the configured URL in the `logger` section of the Seldon Deployment CR. A very simple echo server can be created via the following command:

```
kubectl create -f chapter7/manual_model_deployment/http-
echo-service.yaml -n ml-workshop
```

You should see the following response:

```
$kubectl create -f chapter7/manual_model_deployment/http-echo-service.yaml -n ml-workshop
deployment.apps/logger created
service/logger created
```

Figure 7.27 – Creating a simple HTTP echo server to validate payload logging

Validate that the pod has been created by issuing the following command:

```
kubectl get pods  -n ml-workshop | grep logger
```

You should see the following response:

```
$kubectl get pods  -n ml-workshop | grep logger
logger-848dd74654-8n88z                          1/1      Running
```

Figure 7.28 – Validating a simple HTTP echo server

16. Let's make a call for our model to predict something. The model we developed in the previous chapter is not very useful, but it will help us understand and validate the overall process of packaging and deploying the model.

Recall from *Chapter 6*, *Machine Learning Engineering*, that the `hellomlflow` notebook has the input for the model with shape `(4,2)`, and the output shape is `(4,)`.

```
X = np.array([[1, 1], [1, 2], [2, 2], [2, 3]])
y = np.dot(X, np.array([1, 2])) + 3
```

Figure 7.29 – Input and output for the model

So, if we want to send data to our model, it would be an array of integer pairs such as [2,1]. When you make a call to your model, the input data is required within an ndarray field under a key named data. The input would look as follows. This is the format the Seldon service expects for the data to be sent to it:

```
{
  "data": {
    "ndarray": [
      [2,1]
    ]
  }
}
```

Figure 7.30 – Input for the model as an HTTP payload

17. Next is the REST endpoint for the model. It will be the ingress that you created in *Step 13* and the standard Seldon URL. The final form would be as follows: http://<INGRESS_LOCATION>/api/v1.0/predictions.

This would translate, in my case, to http://model-test.192.168.61.72.nip.io/api/v1.0/predictions.

Now, you have the payload and the URL to send this request to.

18. In this step, you will make a call to your model. We are using a commonly used command-line option to make this call; however, you may choose to use other software, such as Postman, to make this HTTP call.

You will use the POST HTTP verb in the call and then provide the location of the service. You will have to pass the Content-Type header to mention JSON content and the body is passed using the data-raw flag of the curl program:

```
curl -vvvv -X POST 'http://<INGRESS_LOCATION>/api/v1.0/
predictions' \--header 'Content-Type: application/json'
\--data-raw '{  "data": {     "ndarray": [[2,1]]  }}'
```

The final request should look as follows. Before making this call, make sure to change the URL as per your ingress location:

```
curl -vvvv -X POST 'http://model-test.192.168.61.72.
nip.io/api/v1.0/predictions' \--header 'Content-Type:
application/json' \--data-raw '{  "data": {     "ndarray":
[[2,1]]  }}'
```

You should see the following response. Note that the output of the command shows the array of the same shape as per our model, which is (4,), and it is under the `ndarray` key in the following screenshot:

```
{"data":{"names":["t:0","t:1","t:2","t:3"],"ndarray":[[0.25,0.25,0.25,
0.25]]},"meta":{"requestPath":{"model-test-predictor":"quay.io/ml-on-k
8s/hellomlflow-manual:1.0.0"}}}
```

Figure 7.31 – Output payload for the model inference call

19. Now, let's verify that the model payload has been logged onto our echo server. You are validating the capability of Seldon to capture input and output and send it to the desired location for further processing, such as drift detection or audit logging:

```
export LOGGER_POD_NAME=$(kubectl get pod -o=custom-
columns=NAME:.metadata.name -n ml-workshop | grep logger)
kubectl logs -f $LOGGER_POD_NAME -n ml-workshop
```

You will see there is a separate record for the input and the output payload. You can use the `ce-requestid` key to correlate the two records in the logs. The following screenshot displays the main fields of the captured input payload of the inference call:

```
"path": "/",
"headers": {-
},
"method": "POST",
"body": "{\n  \"data\": {\n    \"ndarray\": [[2,1]]\n  }\n}",
"ip": "::ffff:172.17.0.1",
"protocol": "http",
"query": {},
"subdomains": [],
"xhr": false,
"os": {-
},
"connection": {},
"json": {
    "data": {
        "ndarray": [
            [
                2,
                1
            ]
        ]
    }
}
```

Figure 7.32 – Captured input payload forwarded to the echo pod

The following screenshot displays the main fields of the output payload of the inference call:

```
"path": "/",
"headers": {--
},
"connection": {},
"json": {
    "data": {
        "names": [
            "t:0",
            "t:1",
            "t:2",
            "t:3"
        ],
        "ndarray": [
            [
                0.25,
                0.25,
                0.25,
                0.25
            ]
        ]
    },
    "meta": {
        "requestPath": {
            "model-test-predictor": "quay.io/ml-on-k8s/hellomlflow-manual:1.0.0"
        }
    }
}
```

Figure 7.33 – Captured output payload forwarded to the echo pod

20. Now, let's verify that service monitoring data is captured by the Seldon engine and is available for us to use and record. Note that the way Prometheus works is by scraping repetitively, so this data is in the current state and the Prometheus server is responsible for calling this URL and record in its database.

The URL format for this information is as follows. The ingress is the same as you created in *Step 13*:

```
http://<INGRESS_LOCATION>/prometheus
```

This would translate to the following for my ingress:

```
http://model-test.192.168.61.72.nip.io/prometheus
```

Open a browser and access the URL in it. You should see the following response:

```
# HELP go_gc_duration_seconds A summary of the pause duration of
garbage collection cycles.
# TYPE go_gc_duration_seconds summary
go_gc_duration_seconds{quantile="0"} 3.9284e-05
go_gc_duration_seconds{quantile="0.25"} 4.4424e-05
go_gc_duration_seconds{quantile="0.5"} 4.779e-05
go_gc_duration_seconds{quantile="0.75"} 5.3872e-05
go_gc_duration_seconds{quantile="1"} 0.00014359
go_gc_duration_seconds_sum 0.031500144
go_gc_duration_seconds_count 605
# HELP go_goroutines Number of goroutines that currently exist.
# TYPE go_goroutines gauge
go_goroutines 28
# HELP go_info Information about the Go environment.
# TYPE go_info gauge
go_info{version="go1.17.1"} 1
# HELP go_memstats_alloc_bytes Number of bytes allocated and still
in use.
```

Figure 7.34 – Accessing monitoring data in Prometheus format

You will find that a lot of information is captured, including response times, the number of HTTP responses per status code (200, 400, 500, and so on), data capture, server performance, and exposing the Go runtime metrics. We encourage you to go through these parameters to develop an understanding of the data available. In the later chapters, you will see how to harvest and plot this data to visualize the performance of the model inferencing server.

You have done a great deal in this exercise. The aim of this section was to showcase the steps and components involved to deploy a model using Seldon Core. In the next section, you will be introduced to the workflow component of the platform, Airflow, and in the next couple of chapters, all of these steps will be automated using the components in the ML platform.

Introducing Apache Airflow

Apache Airflow is an open source software designed for programmatically authoring, executing, scheduling, and monitoring workflows. A workflow is a sequence of tasks that can include data pipelines, ML workflows, deployment pipelines, and even infrastructure tasks. It was developed by Airbnb as a workflow management system and was later open sourced as a project in Apache Software Foundation's incubation program.

While most workflow engines use XML to define workflows, Airflow uses Python as the core language for defining workflows. The tasks within the workflow are also written in Python.

Airflow has many features, but we will cover only the fundamental bits of Airflow in this book. This section is by no means a detailed guide for Airflow. Our focus is to introduce you to the software components for the ML platform. Let's start with DAG.

Understanding DAG

A workflow can be simply defined as a sequence of **tasks**. In Airflow, the sequence of tasks follows a data structure called a **directed acyclic graph** (**DAG**). If you remember your computer science data structures, a DAG is composed of nodes and one-way vertices organized in a way to ensure that there are no cycles or loops. Hence, a workflow in Airflow is called a DAG.

Figure 7.35 shows a typical example of a data pipeline workflow:

Figure 7.35 – Typical data pipeline workflow

The example workflow in *Figure 7.36* is composed of tasks represented by boxes. The order of execution of these tasks is determined by the direction of the arrows:

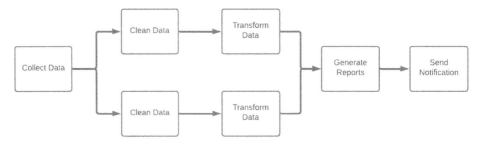

Figure 7.36 – Example workflow with parallel execution

Another example of a workflow is shown in *Figure 7.36*. In this example, there are tasks that are executed in parallel. The **Generate Report** tasks will wait for both **Transform Data** tasks to complete. This is called **execution dependency** and it is one of the problems Airflow is solving. Tasks can only execute if the upstream tasks are completed.

You can configure the workflow however you want as long as there are no cycles in the graph, as shown in *Figure 7.37*:

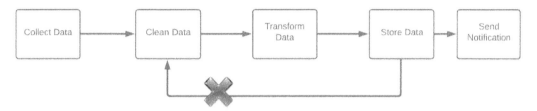

Figure 7.37 – Example workflow with cycle

In the example in *Figure 7.37*, the **Clean Data** task will never be executed because it is dependent on the **Store Data** task, which will also not be executed. Airflow only allows acyclic graphs.

As illustrated, a DAG is a series of tasks, and there are three common types of tasks in Airflow:

- **Operators**: Predefined tasks that you can use to execute something, They can be strung together to form a pipeline or a workflow. Your DAG is composed mostly, if not entirely, of operators.

- **Sensors**: Subtypes of operators that are used for a series of other operators based on an external event.

- **TaskFlow**: Custom Python functions decorated with `@task`. This allows you to run regular Python functions as tasks.

Airflow operators are extendable, which means there are quite a lot of predefined operators created by the community that you can simply use. One of the operators that you will mostly use in the following exercises is the **Notebook Operator**. This operator allows you to run any Jupyter notebook as tasks in the DAG.

So, what are the advantages of using DAGs to execute a sequence of tasks? Isn't it enough to just write a script that can execute other scripts sequentially? Well, the answer lies in the features that Airflow offers, which we will explore next.

Exploring Airflow features

The advantages that Airflow brings when compared with `cron` jobs and scripts can be detailed by its features. Let's start by looking at some of those features:

- **Failure and error management**: In the event of a task failure, Airflow handles errors and failures gracefully. Tasks can be configured to automatically retry when they fail. You can also configure how many times it retries.

  In terms of execution sequence, there are two types of task dependencies in a typical workflow that can be managed in Airflow much easier than writing a script.

- **Data dependencies**: Some tasks may require that the other tasks be processed first because they require data that is generated by other tasks. This can be managed in Airflow. Moreover, Airflow allows the passing of small amounts of metadata from the output of one task as an input to another task.

- **Execution dependencies**: You may be able to script execution dependencies in a small workflow. However, imagine scripting a workflow in Bash with a hundred tasks, where some tasks can run concurrently while others can only run sequentially. I imagine this to be a pretty daunting task. Airflow helps simplify this by creating DAGs.

- **Scalability**: Airflow can horizontally scale to multiple machines or containers. The tasks in the workflow may be executed on different nodes while being orchestrated centrally by a common scheduler.

- **Deployment**: Airflow can use Git to store DAGs. This allows you to continuously deploy new changes to your workflows. A sidecar container can automatically pick up the changes from the `git` repository containing your DAGs. This allows you to implement the continuous integration of DAGs.

The next step is to understand the different components of Airflow.

Understanding Airflow components

Airflow comprises multiple components running as independent services. *Figure 7.38* shows the components of Airflow and their interactions:

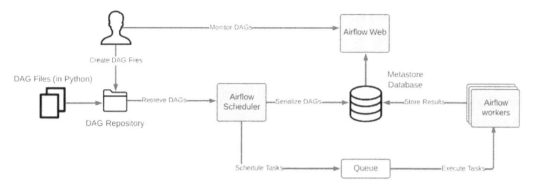

Figure 7.38 – Airflow components

There are three core services in Airflow. The **Airflow Web** serves the user interface where users can visually monitor and interact with DAGs and tasks. The **Airflow Scheduler** is a service responsible for scheduling tasks for the Airflow Worker. Scheduling does not only mean executing tasks according to their scheduled time. It's also about executing the tasks in a particular sequence, taking into account the execution dependencies and failure management. **Airflow Worker** is the service that executes the tasks. This is also the main scalability point of Airflow. The more Airflow Worker is running, the more tasks can be executed concurrently.

The DAG repository is a directory in the filesystem where DAG files written in Python are stored and retrieved by the scheduler. The Airflow instance configured in our platform includes a sidecar container that synchronizes the DAG repository with a remote `git` repository. This simplifies the deployment of DAGs by simply pushing a Python file to Git.

We will not dig too deep into Airflow in this book. The objective is for you to learn enough to a point where you are able to create pipelines in Airflow with minimal Python coding. You will use the Elyra notebooks pipeline builder feature to build Airflow pipelines graphically. If you want to learn more about Airflow and how to build pipelines programmatically in Python, we recommend that you start with Apache Airflow's very rich documentation at `https://airflow.apache.org/docs/apache-airflow/stable/concepts/overview.html`.

Now that you have a basic understanding of Airflow, it's time to take a look at it in action. In *Chapter 4*, *The Anatomy of a Machine Learning Platform*, you installed a fresh instance of ODH. This process also installed the Airflow services for you. Now, let's validate this installation.

Validating the Airflow installation

To validate that Airflow is running correctly in your cluster, you need to perform the following steps:

1. Check whether all the Airflow pods are running by executing the following command:

```
kubectl get pods -n ml-workshop | grep airflow
```

You should see the three Airflow services pods in running status, as shown in the following screenshot in *Figure 7.39*. Verify that all pods are in the Running state:

```
[mlops@fedora ~]$ kubectl get pods -n ml-workshop | grep airflow
app-aflow-airflow-scheduler-f7fc5d4cb-sfzfz    2/2    Running
app-aflow-airflow-web-7c566d79d-cbqsg          2/2    Running
app-aflow-airflow-worker-0                     2/2    Running
```

Figure 7.39 – Airflow pods in the Running state

2. Get the URL of Airflow Web by looking at the ingress host of ap-airflow2. You can do this by executing the following command:

```
kubectl get ingress -n ml-workshop | grep airflow
```

You should see results similar to *Figure 7.39*. Take note of the host value of the ap-airflow2 ingress. The IP address may be different in your environment:

```
[mlops@fedora ~]$ kubectl get ingress -n ml-workshop | grep airflow
ap-airflow2         nginx    airflow.192.168.49.2.nip.io        localhost
```

Figure 7.40 – Airflow ingress in the ml-workshop namespace

3. Navigate to https://airflow.192.168.49.2.nip.io. Note that the domain name is the host value of the ap-airflow2 ingress. You should see the Airflow Web UI, as shown in *Figure 7.41*:

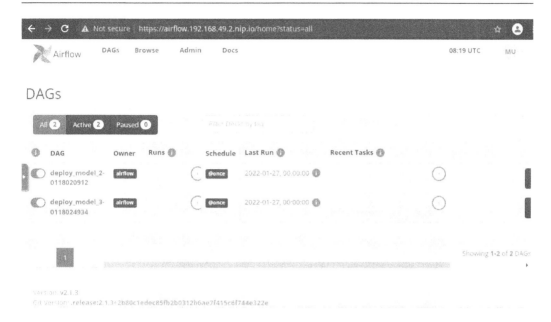

Figure 7.41 – Home screen of Apache Airflow

If you are able to load the Airflow landing page, it means that the Airflow installation is valid. You must have also noticed that in the table listing the DAGs, there are already existing DAGs currently in failing status. These are existing DAG files that are in `https://github.com/airflow-dags/dags/`, the default configured DAG repository. You will need to create your own DAG repository for your experiments. The next section will provide the details on how to do this.

Configuring the Airflow DAG repository

A DAG repository is a Git repository where Airflow picks up the DAG files that represent your pipelines or workflows. To configure Airflow to point to your own DAG repository, you need to create a Git repository and point the Airflow Scheduler and Airflow Web to this Git repository. You will use **GitHub** to create this repository. The following steps will guide you through the process:

1. Create a GitHub repository by going to `https://github.com`. This requires that you have an existing account with GitHub. For the purpose of this exercise, let's call this repository `airflow-dags`. Take note of the URL of your new Git repository. It should look like this: `https://github.com/your-user-name/airflow-dags.git`. We assume that you already know how to create a new repository on GitHub.

2. Edit your instance of ODH by editing the `kfdef` (**Kubeflow definition**) object. You can do this by executing the following command:

```
kubectl edit kfdef opendatahub-ml-workshop -n ml-workshop
```

You should be presented with a `vim` editor showing the `kfdef` manifest file as shown in *Figure 7.42*. Press *i* to start editing.

```
- kustomizeConfig:
    parameters:
    - name: KEYCLOAK_URL
      value: keycloak.192.168.49.2.nip.io
    - name: CLIENT_SECRET
      value: efbb6591-86d0-4270-a544-f47689460a85
    - name: AIRFLOW_HOST
      value: airflow.192.168.49.2.nip.io
    - name: DAG_REPO
      value: https://github.com/airflow-dags/dags.git
    repoRef:
      name: manifests
      path: manifests/airflow2
  name: airflow2
```

Figure 7.42 – vim editor showing the section defining the Airflow instance

3. Replace the value of the `DAG_REPO` parameter with the URL of the Git repository you created in *Step 1*. The edited file should look like the screenshot in *Figure 7.43*. Press *Esc*, then :, and type wq and press *Enter* to save the changes you made to the `kfdef` object.

```
- kustomizeConfig:
    parameters:
    - name: KEYCLOAK_URL
      value: keycloak.192.168.49.2.nip.io
    - name: CLIENT_SECRET
      value: efbb6591-86d0-4270-a544-f47689460a85
    - name: AIRFLOW_HOST
      value: airflow.192.168.49.2.nip.io
    - name: DAG_REPO
      value: https://github.com/your-user-name/airflow-dags.git
    repoRef:
      name: manifests
      path: manifests/airflow2
  name: airflow2
```

Figure 7.43 – Value of the DAG_REPO parameter after editing

The changes will be picked up by the ODH operator and will be applied to the affected Kubernetes deployment objects, in this case, Airflow Web and Airflow Scheduler deployments. This process will take a couple of minutes to complete.

4. Validate the changes by inspecting the Airflow deployments. You can do this by running the following command to look into the applied manifest of the deployment object:

```
kubectl get deployment app-aflow-airflow-scheduler -o
yaml -n ml-workshop | grep value:.*airflow-dags.git
```

This should return a line containing the URL of your GitHub repository.

5. Because this repository is new and is empty, you should not see any DAG files when you open the Airflow Web UI. To validate the Airflow web application, navigate to your Airflow URL, or refresh your existing browser tab, and you should see an empty Airflow DAG list similar to the screenshot in *Figure 7.44*:

Figure 7.44 – Empty Airflow DAG list

Now that you have validated your Airflow installation and updated the DAG repository to your own git repository, it's time to put Airflow to good use.

Configuring Airflow runtime images

Airflow pipelines, or DAGs, can be authored by writing Python files using the Airflow libraries. However, it is also possible to create DAGs graphically from an Elyra notebook. In this section, you will create an Airflow DAG from Elyra, push it to the DAG repository, and execute it in Airflow.

To further validate the Airflow setup and test the configuration, you will need to run a simple `Hello world` pipeline. Follow the steps to create a two-task pipeline. You will create Python files, a pipeline, and configure runtime images to be used throughout the process:

1. If you do not have a running notebook environment, start a notebook environment by navigating to JupyterHub, clicking **Start My Server**, and selecting a notebook image to run, as shown in *Figure 7.45*. Let's use **Base Elyra Notebook Image** this time as we do not require any special libraries.

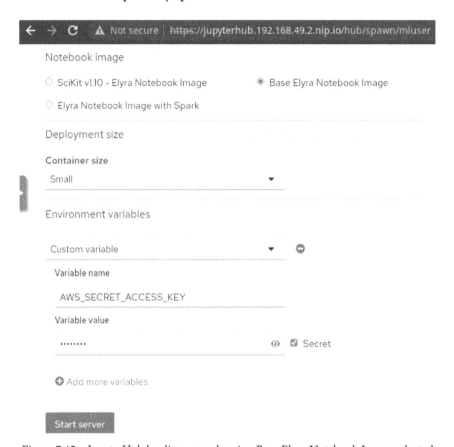

Figure 7.45 – JupyterHub landing page showing Base Elyra Notebook Image selected

2. In your Elyra browser, navigate to the `Machine-Learning-on-Kubernetes/` `chapter7/model_deploy_pipeline/` directory.

3. Open a new pipeline editor. You can do this by selecting the menu item **File>New>Pipeline Editor**, as shown in *Figure 7.46*. A new file will appear in the left-hand browser, named `untitled.pipeline`.

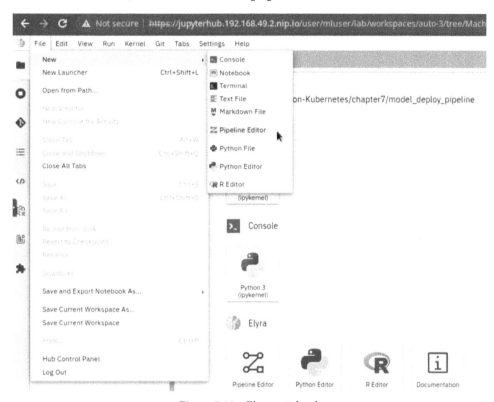

Figure 7.46 – Elyra notebook

4. Right-click on the `untitled.pipeline` file and rename it to `hello_world.` `pipeline`.

5. Create two Python files with the same contents containing the following line: `print('Hello airflow!')`. You can do this by selecting the menu items **File** > **New Python** File. Then, rename the files to `hello.py` and `world.py`. Your directory structure should look like the screenshot in *Figure 7.47*:

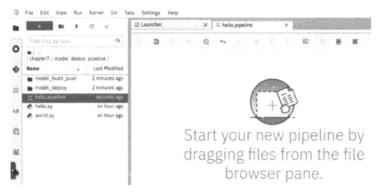

Figure 7.47 – Elyra directory structure showing the hello.pipeline file

6. Create a pipeline with two tasks by dragging the `hello.py` file into the pipeline editor window. Do the same for `world.py`. Connect the tasks by dragging the tiny circle on the right of the task box to another box. The resulting pipeline topology should look like the illustration in *Figure 7.48*. Save the pipeline by clicking the **Save** icon in the top toolbar.

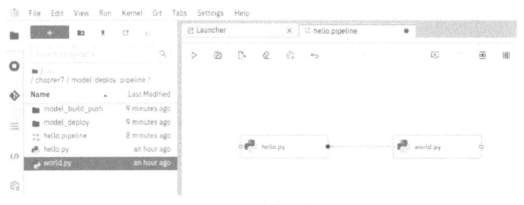

Figure 7.48 – Task topology

7. Before we can run this pipeline, we need to configure each of the tasks. Because each task will run as a container in Kubernetes, we need to tell which container image that task will use. Select the **Runtime Images** icon on the toolbar on the left. Then, click the + button to add a new runtime image, as shown in *Figure 7.49*:

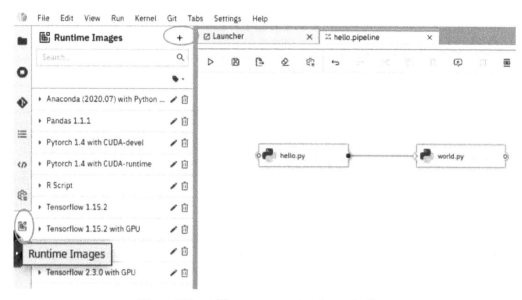

Figure 7.49 – Adding a new runtime image in Elyra

8. In the **Add new Runtime Image** dialog, add the details of the **Kaniko Container Builder** image, as shown in *Figure 7.50*, and hit the **SAVE & CLOSE** button.

This container image (`https://quay.io/repository/ml-on-k8s/kaniko-container-builder`) contains the tools required to build Docker files and push images to an image registry from within Kubernetes. This image can also pull ML models and metadata from the MLflow model registry. You will use this image to build containers that host your ML model in the next section. This container image was created for the purpose of this book. You can use any container image as a runtime image for your pipeline tasks as long as the image can run on Kubernetes.

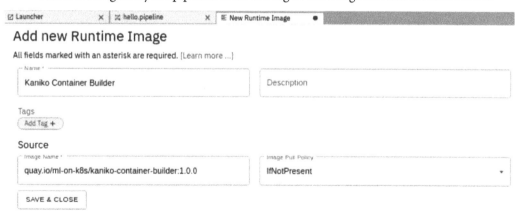

Figure 7.50 – Add new Runtime Image dialog for Kaniko builder

9. Add another runtime image called **Airflow Python Runner**. The container image is located at `https://quay.io/repository/ml-on-k8s/airflow-python-runner`. This image can run any Python 3.8 scripts, and interact with Kubernetes and Spark operators. You will use this image to deploy container images to Kubernetes in the next section. Refer to *Figure 7.51* for the **Add new Runtime Image** dialog field values, and then hit the **SAVE & CLOSE** button:

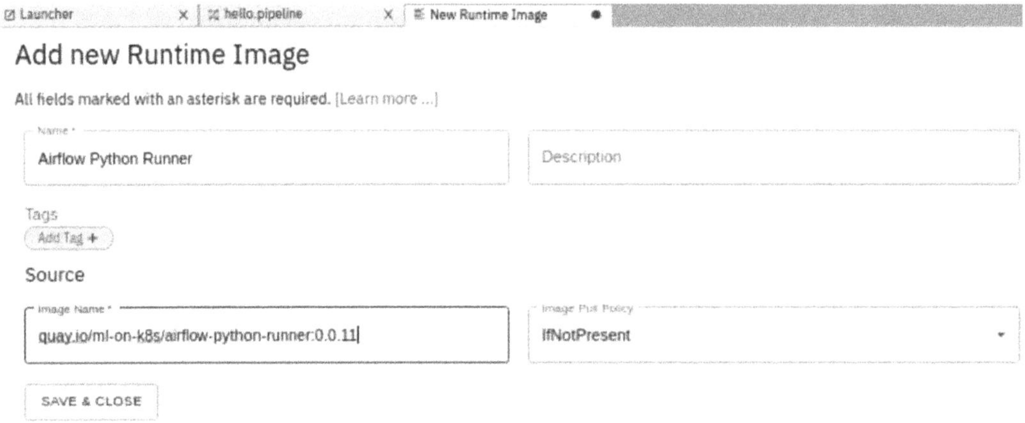

Figure 7.51 – Add new Runtime Image dialog for Airflow Python Runner

10. Pull the images from the remote repository to the local Docker daemon of your Kubernetes cluster. This will help speed up the start up times of tasks in Airflow by using a runtime image that is already pulled into the local Docker instance.

    You can do this by running the following command on the same machine where your Minikube is running. This command allows you to connect your Docker client to the Docker daemon inside your Minikube **virtual machine (VM)**:

    ```
    eval $(minikube docker-env)
    ```

11. Pull the **Kaniko Container Builder** image by running the following command in the same machine where your Minikube is running. This will pull the image from `quay.io` to the Docker daemon inside your Minikube:

    ```
    docker pull quay.io/ml-on-k8s/kaniko-container-
    builder:1.0.0
    ```

12. Pull the **Airflow Python Runner** image by running the following command in the same machine where your Minikube is running:

    ```
    docker pull quay.io/ml-on-k8s/airflow-python-
    runner:0.0.11
    ```

13. Assign **Kaniko Container Builder** runtime images to the `hello.py` task. You can do this by right-clicking the task box and selecting the **Properties** context menu item. The properties of the task will be displayed in the right pane of the pipeline editor, as shown in *Figure 7.52*. Using the **Runtime Image** drop-down box, select **Kaniko Container Builder**.

Figure 7.52 – Setting the runtime image of a task in the pipeline editor

> **Note**
>
> If you do not see the newly added runtime images in the drop-down list, you need to close and reopen the pipeline editor. This will refresh the list of runtime images.

14. Assign the **Airflow Python Runner** runtime image to the `world.py` task. This is similar to *Step 10*, but for the `world.py` task. Refer to *Figure 7.53* for the **Runtime Image** value:

Figure 7.53 – Setting the runtime image of a task in the pipeline editor

15. You have just created an Airflow pipeline that has two tasks, where each task uses a different runtime. But, before we can run this pipeline in Airflow, we need to tell Elyra where Airflow is. To do this, select the **Runtimes** icon on the left toolbar of Elyra, as shown in *Figure 7.54*:

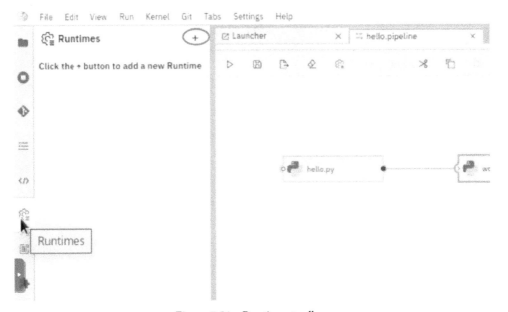

Figure 7.54 – Runtimes toolbar

16. Hit the + button and select the **New Apache Airflow runtime** menu item. Fill in the details according to the following values or see *Figure 7.55*:

A. **Apache Airflow UI Endpoint** is where the Airflow UI is currently listening. This is not critical, as Elyra does not interact with Airflow UI directly. Set the value to the URL of your Airflow UI. This will look like `https://airflow.192.168.49.2.nip.io`, where the IP address part is the IP address of your Minikube.

B. **Apache Airflow User Namespace** is the Kubernetes namespace where all the pods of the tasks will be created. Set this to `ml-workshop`. This is the namespace of all your ML platform workloads.

C. **GitHub DAG Repository** is the DAG repository that you created in the previous section, *Configuring Airflow DAG Repository*. This follows the `github-username/airflow-dags` format. Replace `github-username` with your GitHub username.

D. **GitHub DAG Repository Branch** is the branch in your GitHub repository where Elyra will push the DAG files. Set this to **main**.

E. **GitHub Personal Access Token** is your GitHub user token with permission to push to your DAG repository. You can refer to the GitHub documentation for creating personal access tokens at `https://docs.github.com/en/ authentication/keeping-your-account-and-data-secure/ creating-a-personal-access-token`.

F. **Cloud Object Storage Endpoint** is the endpoint URL of any S3 storage API. Airflow uses this to publish artifacts and logs of the DAG executions. You will use the same Minio server for this. Set the value to `http://minio-ml- workshop:900`. This is the URL of the Minio service. We did not use the Minio's ingress because the JupyterHub server is running on the same Kubernetes namespace as the Minio server, which means that the Minio service can be addressed by its name.

G. **Cloud Object Storage User name** is the Minio username, which is `minio`.

H. **Cloud Object Storage Password** is the Minio password, which is `minio123`.

Once all the fields are filled correctly, hit the **SAVE & CLOSE** button.

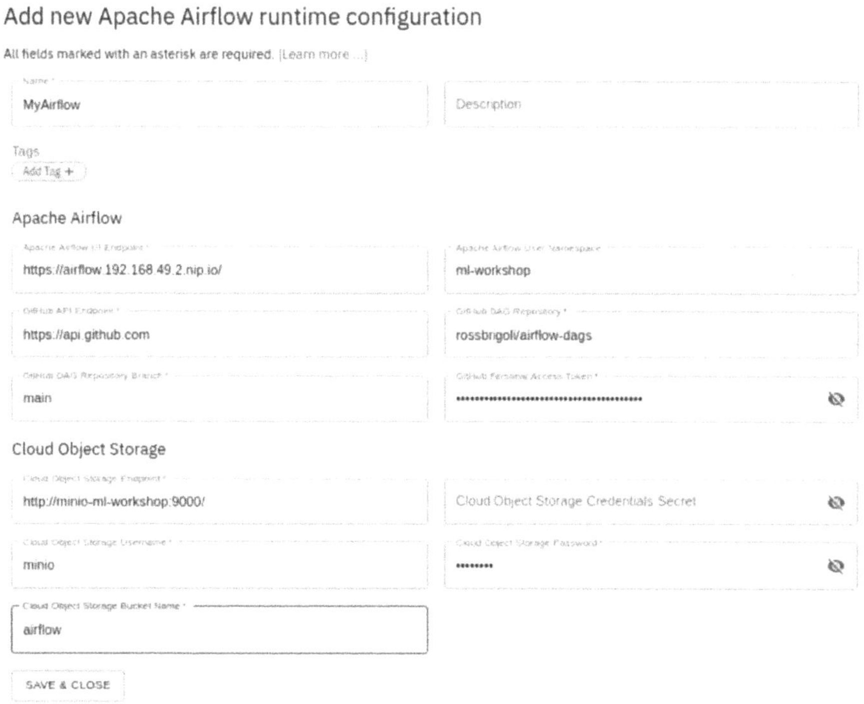

Add new Apache Airflow runtime configuration

All fields marked with an asterisk are required. [Learn more ...]

Name *
MyAirflow

Description

Tags
Add Tag +

Apache Airflow

Apache Airflow UI Endpoint *
https://airflow.192.168.49.2.nip.io/

Apache Airflow User Namespace
ml-workshop

GitHub API Endpoint *
https://api.github.com

GitHub DAG Repository *
rossbrigoli/airflow-dags

GitHub DAG Repository Branch *
main

GitHub Personal Access Token *
•••

Cloud Object Storage

Cloud Object Storage Endpoint *
http://minio-ml-workshop:9000/

Cloud Object Storage Credentials Secret

Cloud Object Storage Username *
minio

Cloud Object Storage Password *
••••••••

Cloud Object Storage Bucket Name *
airflow

SAVE & CLOSE

Figure 7.55 – Adding a new Apache Airflow runtime configuration

17. Run the pipeline in Airflow by clicking the **Play** button in the top toolbar of the pipeline editor. This will bring up a **Run pipeline** dialog. Select **Apache Airflow runtime** as the runtime platform and **MyAirflow** as the runtime configuration, and then hit **OK**. Refer to *Figure 7.56*:

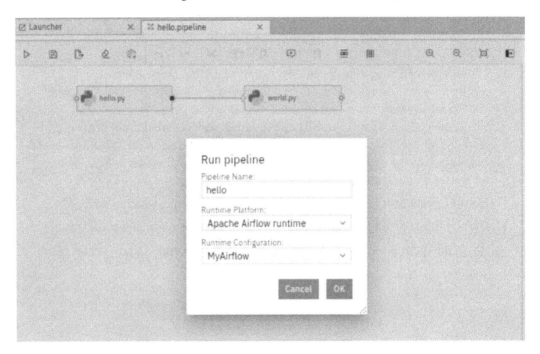

Figure 7.56 – Run pipeline dialog

This action generates an Airflow DAG file and pushes the file to the GitHub repository configured as a DAG repository. You can verify this by checking your GitHub repository for newly pushed files.

18. Open the Airflow website. You should see the newly create DAG, as shown in *Figure 7.57*. If you do not see it, refresh the Airflow page a few times. Sometimes, it takes a few seconds before the DAGs appear in the UI.

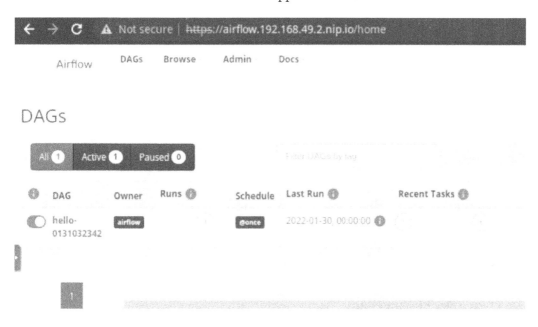

Figure 7.57 – Airflow showing a running DAG

The DAG should succeed in a few minutes. If it does fail, you need to review the steps to make sure you set the correct values and that you did not miss any steps.

You have just created a basic Airflow DAG using Elyra's graphical pipeline editor. The generated DAG is, by default, configured to only run once, indicated by the @once annotation. In the real world, you may not want to run your DAGs directly from Elyra. You may want to add additional customizations to the DAG file. In this case, instead of running the DAG by clicking the play button, use the export feature. This will export the pipeline into a DAG file that you can further customize, such as setting the schedule. You can then push the customized DAG file to the DAG repository to submit it to Airflow.

You have just validated your Airflow setup, added Airflow runtime configuration, and integrated Elyra with Airflow. Now it is time to build a real deployment pipeline!

Automating ML model deployments in Airflow

You have seen in the preceding sections how to manually package an ML model into a running HTTP service on Kubernetes. You have also seen how to create and run basic pipelines in Airflow. In this section, you will put this new knowledge together by creating an Airflow DAG to automate the model deployment process. You will create a simple Airflow pipeline for packaging and deploying an ML model from the MLflow model registry to Kubernetes.

Creating the pipeline by using the pipeline editor

Similar to the previous section, you will use Elyra's pipeline editor to create the model build and deployment DAG:

1. If you do not have a running Elyra environment, start a notebook environment by navigating to JupyterHub, clicking **Start My Server**, and selecting a notebook image to run, as shown in *Figure 7.45*. Let's use **Base Elyra Notebook Image** because this time, we do not require any special libraries.

2. In your Elyra browser, navigate to the `Machine-Learning-on-Kubernetes/chapter7/model_deploy_pipeline/` directory.

3. Open a new pipeline editor. You can do this by selecting the menu item **File>New>Pipeline Editor**, as shown in *Figure 7.46*. A new file will appear in the left-hand browser, named `untitled.pipeline`.

4. Right-click on the `untitled.pipeline` file and rename it `model_deploy.pipeline`. Your directory structure should look like the screenshot in *Figure 7.58*:

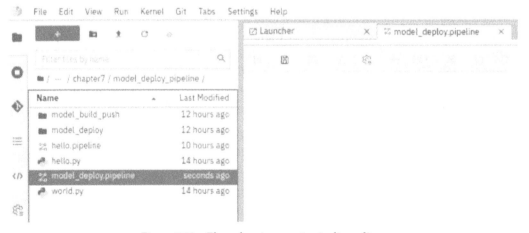

Figure 7.58 – Elyra showing empty pipeline editor

5. You will build a pipeline with two tasks in it. The first task will pull the model artifacts from the MLflow model registry, package the model as a container using Seldon core, and then push the container image to an image repository. To create the first task, drag and drop the `build_push_image.py` file from the `model_build_push` directory to the pipeline editor's workspace. This action will create a new task in the pipeline editor window, as shown in *Figure 7.59*:

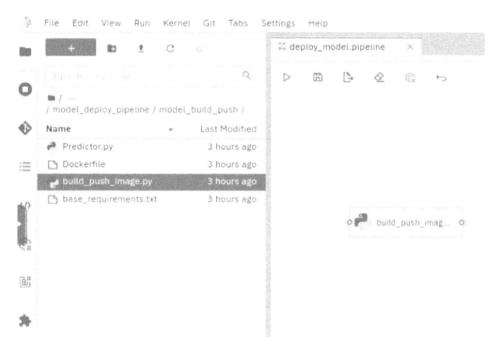

Figure 7.59 – Elyra pipeline editor showing the build_push_image task

6. The second task will pull the container image from the image repository and deploy it to Kubernetes. Create the second task by dragging the `deploy_model.py` file from `model_deploy directory` and dropping it into the pipeline editor workspace. This action will create a second task in the pipeline editor, as shown in *Figure 7.60*:

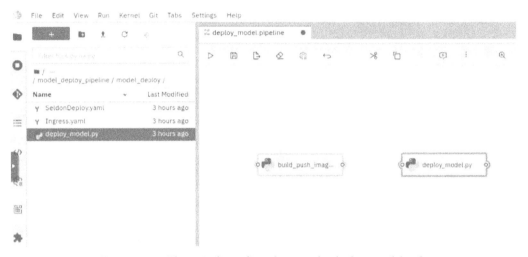

Figure 7.60 – Elyra pipeline editor showing the deploy_model task

7. Connect the two tasks by dragging the tiny circle at the right-hand side of the `build_push_image.py` task to the `deploy_model.py` task box. The task topology should look like the illustration in *Figure 7.61*. Take note of the direction of the arrow highlighted in the red box.

Figure 7.61 – Task topology of the DAG

8. Configure the build_push_image.py task by right-clicking the box and selecting **Properties**. A property panel will appear on the right side of the editor, as shown in *Figure 7.62*. Select **Kaniko Container Builder** as the runtime image for this task.

Figure 7.62 – Pipeline editor with the property panel displayed showing the Kaniko Builder runtime

9. Add file dependencies to build_push_image.py by clicking the **Add Dependency** button and selecting the following files. The file dependencies for this task are also shown in *Figure 7.62*. The following list describes what each file does:

    - Dockerfile – This is the Docker file that will be built to produce the container image that contains the ML model and the Predictor Python file.

    - Predictor.py – This is the Python file used by Seldon to define the inference graph. You have seen this file in the preceding section.

    - Base_requirements.txt – This is a regular text file that contains a list of Python packages required to run this model. This is used by the pip install command inside the Docker file.

10. At this point, you should have an idea of what the entire pipeline does. Because the pipeline needs to push a container image to a registry, you will need a container registry to hold your ML model containers. Create a new repository in a container registry of your choice. For the exercises in this book, we will use **Docker Hub** as an example. We assume that you know how to create a new repository in `https://hub.docker.com`. Call this new repository `mlflowdemo`.

11. Once you have the image repository created, set the **Environment Variables** for the `build_push_image.py` task, as shown in *Figure 7.63*. The following are the six variables you need to set:

  - `MODEL_NAME` is the name of the ML model registered in MLflow. You used the name `mlflowdemo` in the previous sections. Set the value of this variable to `mlflowdemo`.

  - `MODEL_VERSION` is the version number of the ML model registered in MLflow. Set the value of this variable to `1`.

  - `CONTAINER_REGISTRY` is the container registry API endpoint. For Docker Hub, this is available at `https://index.docker.io/v1`. Set the value of this variable to `https://index.docker.io/v1/`.

  - `CONTAINER_REGISTRY_USER` is the username of the user who will push images to the image registry. Set this to your Docker Hub username.

  - `CONTAINER_REGISTRY_PASSWORD` is the password of your Docker Hub user. In production, you do not want to do this. You may use secret management tools to serve your Docker Hub password. However, to keep things simple for this exercise, you will put your Docker Hub password as an environment variable.

  - `CONTAINER_DETAILS` is the name of the repository where the image will be pushed, along with the name and tag of the image. This includes the Docker Hub username in the `your-username/mlflowdemo:latestv` format.

Save the changes by clicking the **Save** icon from the top toolbar of the pipeline editor:

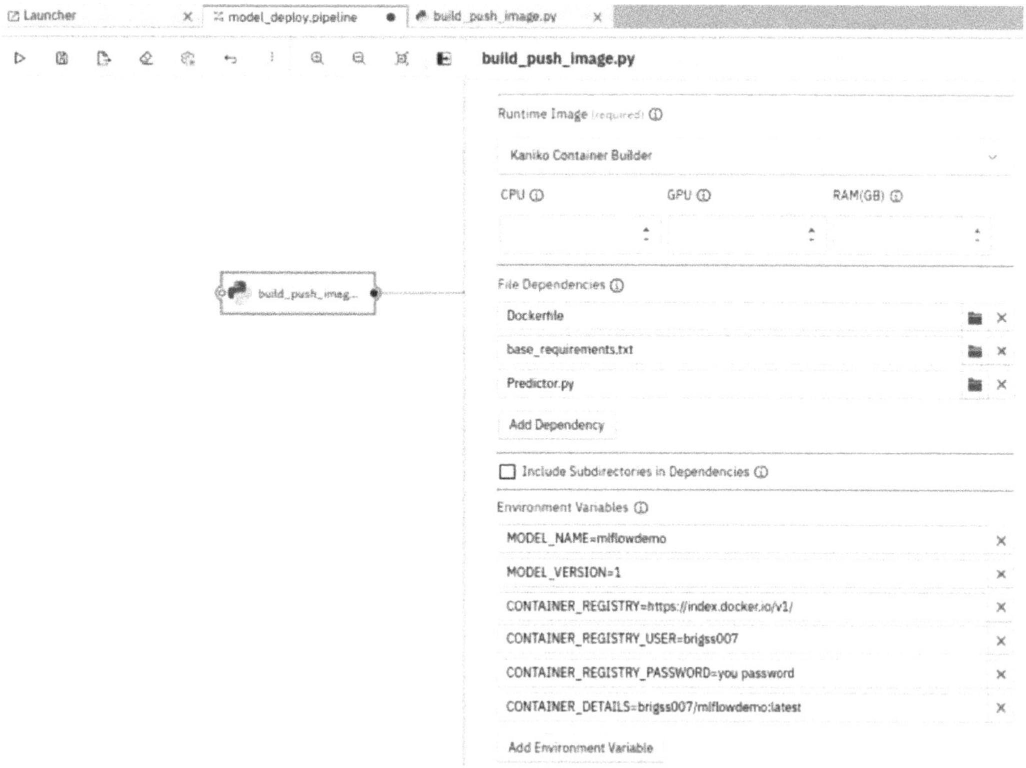

Figure 7.63 – Example environment variables of the build_push_image.py task

12. Configure the `deploy_model.py` task by setting the runtime image, the file dependencies, and the environment variables, as shown in *Figure 7.64*. There are four environment variables you need to set, as detailed in the following list:

A. `MODEL_NAME` is the name of the ML model registered in MLflow. You used the name `mlflowdemo` in the previous sections. Set the value of this variable to `mlflowdemo`.

B. `MODEL_VERSION` is the version number of the ML model registered in MLflow. Set the value of this variable to `1`.

C. `CONTAINER_DETAILS` is the name of the repository to where the image will be pushed and the image name and tag. This includes the Docker Hub username in the `your-username/mlflowdemo:latest` format.

D. CLUSTER_DOMAIN_NAME is the DNS name of your Kubernetes cluster, in this case, the IP address of Minikube, which is <Minikube IP>.nip.io. For example, if the response of the minikube ip command is 192.168.49.2, then the cluster domain name is 192.168.49.2.nip.io. This is used to configure the ingress of the ML model HTTP service so that it is accessible outside the Kubernetes cluster.

Save the changes by clicking the **Save** icon from the top toolbar of the pipeline editor.

Figure 7.64 – Properties of the deploy_model.py task

13. You are now ready to run the pipeline. Hit the **Play** button from the top toolbar of the pipeline editor. This will bring up the **Run** pipeline dialog, as shown in *Figure 7.65*. Select **Apache Airflow runtime** under **Runtime Platform**, and **MyAirflow** under **Runtime Configuration**. Click the **OK** button. This will generate the Airflow DAG Python file and push it to the Git repository.

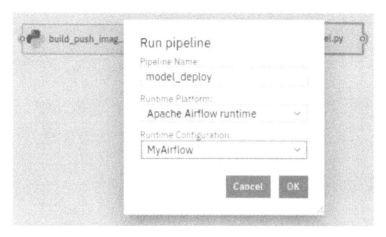

Figure 7.65 – Run pipeline dialog

14. Once the DAG is successfully generated and pushed to the `git` repository, you should see a dialog as shown in *Figure 7.66*. Click **OK**.

Figure 7.66 – DAG submission confirmation dialog

15. Navigate to Airflow's GUI. You should see a new DAG, labeled **model_deploy-some-number**, appear in the DAGs table, and it should start running shortly, as shown in *Figure 7.67*. The mint green color of the job indicates that it is currently running. Dark green indicates that it is successful.

> **Note**
> If you do not see the new DAG, refresh the page until you see it. It may take a few seconds for the Airflow to sync with the Git repository.

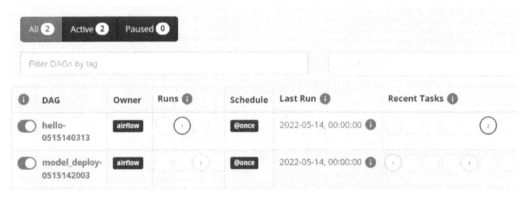

Figure 7.67 – Airflow GUI showing the model_deploy DAG

16. Meanwhile, you can explore the DAG by clicking the DAG name and selecting the **Graph View** tab. It should display the topology of tasks as you designed it in Elyra's pipeline editor, as shown in *Figure 7.68*. You may explore the DAG further by selecting the <> **Code** tab. This will display the generated source code of the DAG.

Figure 7.68 – Graph view of the model_deploy DAG in Airflow

17. After a few minutes, the job should succeed and you should see the outline of all the tasks in **Graph View** turn to dark green. You can also explore the tasks by looking at the pods in Kubernetes. Run the following command and you should see two pods with the **Completed** status, as shown in *Figure 7.69*. These pods are the two tasks in the pipeline that have been executed successfully:

```
kubectl get pods -n ml-workshop\
```

You should see the following response:

```
[mlops@fedora ~]$ k get pods -n ml-workshop
NAME                                                         READY  STATUS    RESTARTS       AGE
app-aflow-airflow-scheduler-6c79785d7d-kr4gm                 2/2    Running   12 (40h ago)   4d1h
app-aflow-airflow-web-58f448f4b6-zngxs                       2/2    Running   12 (16h ago)   4d1h
app-aflow-airflow-worker-0                                   2/2    Running   22 (175m ago)  2d21h
app-aflow-postgresql-0                                       1/1    Running   5 (5d3h ago)   25d
app-aflow-redis-master-0                                     1/1    Running   8 (5d3h ago)   26d
build-push-image.5144cadff01742d78370ef177d6733c6            0/1    Completed 0              18m
deploy-model.a4d75a4d859046bd89566983e2ac49b3                0/1    Completed 0              12m
grafana-5dc6cf89d-mb68x                                      1/1    Running   8 (5d3h ago)   26d
jupyterhub-7848ccd4b7-f9wbw                                  1/1    Running   5 (5d3h ago)   25d
jupyterhub-db-0                                              1/1    Running   2 (5d3h ago)   5d8h
jupyterhub-nb-mluser                                         1/1    Running   0              3d20h
logger-848dd74654-hfx9g                                      1/1    Running   0              3d21h
minio-ml-workshop-6b84fdc7c4-2n9zl                           1/1    Running   0              2d2h
mlflow-d65ccb65d-65ts6                                       2/2    Running   39 (5d3h ago)  26d
mlflow-db-0                                                  1/1    Running   7 (5d3h ago)   26d
model-1-mlflowdemo-predictor-0-predictor-7fb4f86bcc-cw42b    2/2    Running   0              42h
model-test-predictor-0-model-test-predictor-7dc869bcb8-qs8w2 2/2    Running   0              3d21h
prometheus-operator-58cccbff56-7pklc                         1/1    Running   9 (5d3h ago)   26d
seldon-controller-manager-7f67f4985b-8dn42                   1/1    Running   21 (2d3h ago)  26d
spark-operator-77fdf56dbc-9mk9f                              1/1    Running   5 (5d3h ago)   25d
```

Figure 7.69 – Kubernetes pods with a Completed status

You have just created a complete ML model build and deployment pipeline using Seldon Core, Elyra's pipeline editor, orchestrated by Airflow, and deployed to Kubernetes.

Seldon Core and Airflow are big tools that have a lot more features that we have not covered and will not be entirely covered in this book. We have given you the essential knowledge and skills to start exploring these tools further as part of your ML platform.

Summary

Congratulations! You made it this far!

As of this point, you have already seen and used JupyterHub, Elyra, Apache Spark, MLflow, Apache Airflow, Seldon Core, and Kubernetes. You have learned how these tools can solve the problems that MLOps is trying to solve. And, you have seen all these tools running well on Kubernetes.

There are a lot more things that we want to show you on the platform. However, we can only write so much, as the features of each of those tools that you have seen are enough to fill an entire book.

In the next chapter, we will take a step back to look at the big picture of what has been built so far. Then, you will start using the platform end-to-end on an example use case. You will be wearing different hats, such as data scientist, ML engineer, data engineer, and a DevOps person in the succeeding chapters.

Part 3: How to Use the MLOps Platform and Build a Full End-to-End Project Using the New Platform

This section will show you how to build a complete ML project using the platform built in the last section. The chapters in this section will put our platform to the test. This section will define a complete ML life cycle and then will process the data, and build and deploy the model using the platform.

This section comprises the following chapters:

- *Chapter 8, Building a Complete ML Project Using the Platform*
- *Chapter 9, Building Your Data Pipeline*
- *Chapter 10, Building, Deploying, and Monitoring Your Model*
- *Chapter 11, Machine Learning on Kubernetes*

8

Building a Complete ML Project Using the Platform

Until now, you have seen a few components of the platform and how it works. You will start this chapter by understanding the platform at a macro level. The holistic view will help you see how the components weave a complete solution for your **machine learning (ML)** needs.

In the later part of this chapter, you will see how you can start an ML project by using a simple example and how the teams and platform will help achieve your required goal.

In this chapter, you will learn about the following topics:

- Reviewing the complete picture of the ML platform
- Understanding the business problem
- Data collection, processing, and cleaning
- Performing exploratory data analysis

- Understanding feature engineering
- Building and evaluating the ML model
- Reproducibility

Reviewing the complete picture of the ML platform

In the preceding chapters, you have built a complete ML platform on top of Kubernetes. You have installed, configured, and explored the different components of the platform. Before you start using the platform, let's take a step back and look at the platform you have built from the tooling perspective. *Figure 8.1* shows the complete logical architecture of the platform:

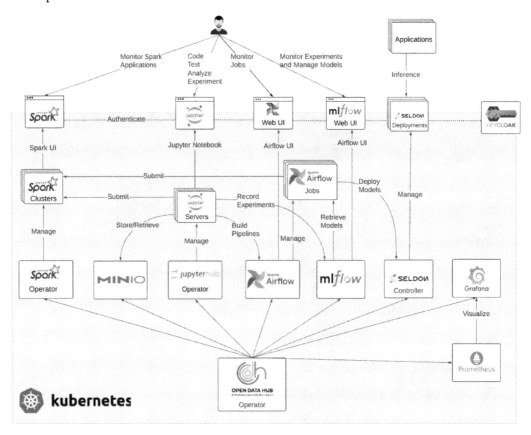

Figure 8.1 – Logical platform architecture

The diagram in *Figure 8.1* also shows the interaction of each platform component. The entire platform runs inside Kubernetes and is managed entirely by the **Open Data Hub (ODH)** operator. Although MinIO is not officially part of the ODH project, we have added it as another component operated by the ODH operator on the platform you just built. In the real world, you will have an S3 storage server already in place, and you will not need to include MinIO in your Kfdef file. It is also important to note that the ODH operator allows you to add or remove tools or swap one tool for another. For example, you could use Argo CD for model deployments instead of Airflow. Keycloak is also not part of the ODH project. However, the components must be secured by a single sign-on mechanism, and Keycloak is one of the best open source tools that can be used to add a single sign-on capability to the platform.

Starting at the top of the diagram, you can see that end users interact with Jupyter notebooks, and the Spark, Airflow, and MLflow UIs. You have seen and experienced these interactions in the preceding chapters. The deployed ML model can then be used for inferencing by applications through REST API calls.

In the middle of the diagram, you can see the interactions between the components and the kind of interactions they perform with each other. Jupyter servers and Airflow jobs can submit Spark applications to the managed Spark clusters. Airflow interacts with the MLflow model registry, while Jupyter notebooks can interact with MLflow to record experiment runs. Airflow also creates Seldon deployment objects that the Seldon controller then converts into running Pods with ML models exposed as REST services. There is no limit to how one component can interact with other platform components.

At the bottom of the diagram, the ODH operator manages and operates the platform components. The ODH operator handles the installation and updates of these components. Spark, JupyterHub, and the Seldon controller are also Kubernetes operators that manage instances of Spark clusters, Jupyter notebook servers, and Seldon deployments, respectively.

Lastly, the ODH operator also manages the Prometheus and Grafana instances. Prometheus is used to collect metrics from each of the components, including the statistics of Seldon deployments. Grafana can then visualize those metrics and can be configured to raise alerts.

The ODH project is still evolving. There may be changes as to what components will be included or excluded in the project in the future. Some of the officially supported components may get replaced with another over time. Therefore, it is important to understand the architecture and how the ODH operator works so that you keep it up to date.

In the next sections, we will take a step back and understand ML projects a bit more, starting with identifying opportunities where an ML solution fits. You will be taken through a scenario that will lead to the creation of a complete ML project.

Understanding the business problem

As with any software project, the first thing is to agree on the business problem you are trying to solve. We have chosen a fictitious scenario for this book to keep it simple while focusing on the process. You can apply the same approach to more complex projects.

Let's assume that you work for an airline booking company as a lead data analyst. The business team of your company has reported that lots of customers complain about flights being delayed. It is causing the company to have bad customer experiences, and the phone staff spend lots of time explaining the details to customers. The business is looking at you to provide a solution to identify which airlines and which flights and times have a lower probability of delays so that the website can prioritize those airlines and, therefore, customers end up with fewer delays.

Let's take a breather here and analyze how we can solve this problem. Do we need ML here? If we take the historical data and place the airlines into two buckets of *delayed* and *on time*, with each bucket placing the airlines into the right category, this attribute can then be used while the customer searches for airlines with better on-time performance. A team of data analysts will analyze the data and assign the ratings. Job done!

While exploring this set, the business has mentioned that one bucket per airline may not provide the granularity that the solution requires. They would like to assess the performance, not at the airline level, but using other factors such as origin and destination airport, and time of day. So, airline A, with flights from Sydney to Melbourne, may go into the *on time* bucket, while the same airline may go into the *delayed* bucket when flying from Tokyo to Osaka. This suddenly expands the scope of the problem. If you need to analyze data at this granularity, it will take a lot of time to process and assign the correct category, and you may need to analyze this data very frequently.

Have you started to think about how you can automate this? The business then mentions that the weather plays a vital role in this problem, and the forecast data from the weather bureau will need to be fetched and preprocessed to perform the analysis. You realize that performing this job with human teams will be slow and complicated and does not provide the solution that the business is looking for. You then mention to the business that you will need to investigate the existing data, which can be used to predict the correct category for a particular flight. You and the business agree that the aim is to predict the flight delay 10 days before the scheduled time with at least 75% accuracy, to improve the customer experience. You will also discuss the response time requirements for the model and understand how the model will be used in the overall business process.

You have just defined the success criteria of this project. You have conveyed to the business that your team will analyze available data to assess its suitability for the project and then plan for the next steps. You have asked the business to associate a **subject matter expert** (**SME**) who can assist in data exploration at this stage.

To summarize, you have outlined the business objectives and the scope of the project. You have also defined the evaluation criteria through which the success of the project would be measured. It is critical that you keep a note of the business value through each stage of the ML life cycle.

Once you have defined the criteria, the next step is to start looking at the available data. For this use case, the data is available at `https://www.kaggle.com/usdot/flight-delays?select=flights.csv`.

Data collection, processing, and cleaning

In this stage, you will begin with gathering raw data from the identified sources. You will write data pipelines to prepare and clean the raw data for analysis.

Understanding data sources, location, and the format

You have started working with the SME to access a subset of the flight data. You will understand the data format and the integration process required to access this data. The data could be in CSV format, or it may be available in some **relational database management system** (**RDBMS**). It is vital to understand how this data would be available for your project and how this data is being maintained eventually.

Start this process by identifying what data is easily available. The SME has mentioned that the flight records data that covered the flight information, the scheduled and actual departure times, and the scheduled and actual arrival times is readily available. This information is available in the object store of your organization. This could be a good starting point.

Understanding data processing and cleaning

The data collected from the raw data sources may have many problems. The collected data may have duplication, missing values, and/or invalid records. For example, you may find that a column of the `string` type may have numerical data in it. You will then work with the SME to find out ways to handle the anomalies.

How would you handle the missing data? Choose an estimated value of the missing data from the existing set. Or you may decide to drop the column altogether if there are many missing values and you can not find any way to impute the missing values.

Implement data validation checks to make sure that the cleaned dataset has consistency and that the data quality problems described here are properly handled. Imagine that the age column has a value of 250. Although we would all like to live this long or beyond, clearly this data is not valid. During this stage, you will find the discrepancy in the data and work out how to handle it.

You may find that the flight arrival and departure times are in the local time zones, and you may choose to add a new column with the times represented in UTC format for easier comparisons.

Data cleaning can happen in both the data engineering stage and the model development stage. Data anomalies that are related to the domain or business logic may be found and handled in the data engineering stage, while data augmentation and data encoding are done at the model development stage. This is because it is the data scientist or the ML engineer who knows best what data formats the model training requires, while the data engineers work closer to the business domain experts.

One way to implement such data validation in the data engineering phase is through Apache Spark. Spark has a set of built-in functions that you can use for data cleaning. The following code shows an example of how to filter out invalid rows or rows that contain malformed data while reading from a data source:

```
dataframe = spark.read.option("header", True).option("mode",
'DROPMALFORMED').csv('flights.csv')
```

Another example is the `fillna()` function. It is used to replace null values with any other values. The following example shows how to replace all null values in the data frame with zeros:

```
dataframe = dataframe.fillna(value=0)
```

On the model development side, there are several techniques to perform the same operations using pandas to manipulate data frames. You will see this in action in the following chapters.

Once you have executed the data cleaning pipeline and created an intermediary dataset that can be used for the next stage, the next step is to see whether the available data helps you in achieving the business goal.

Performing exploratory data analysis

At this stage, you analyze the data to assess its suitability for the given problem. Data analysis is essential for building ML models. Before you create an ML model, you need to understand the context of the data. Analyzing vast amounts of company data and converting it into a useful result is extremely difficult, and there is no single answer on how to do it. Figuring out what data is meaningful and what data is vital for business is the foundation for your ML model.

This is a preliminary analysis, and it does not guarantee that the model will bring the expected results. However, it provides an opportunity to understand the data at a higher level and pivot if required.

Understanding sample data

When you get a set of data, you first try to understand it by merely looking at it. You then go through the business problem and try to determine what set of patterns would be helpful for the given situation. A lot of the time, you will need to collaborate with SMEs who have relevant domain knowledge.

At this stage, you may choose to convert the data into a tabular form to better understand it. Classify the columns according to the data values. Understand each variable in the dataset and find out whether the values are continuous, or whether it represents a category. You will then summarize the columns using descriptive statistics to understand the values your columns contain. These statistics could be mean or median or anything that helps you understand the data.

Understand the **data variance**. For example, your data has only 5% records of delayed flights and the remaining flights are on time. Would this dataset be good for your desired outcomes? You need to get a better dataset that represents a more balanced distribution. You may choose to downsample the dataset, if it is highly imbalanced, by reducing the examples from the majority class.

Humans are good at visualizing data so, to better understand the data, you will need to visualize your columns using charts. There is a series of different charts that can help you visualize your data. The platform we present here will assist you in writing code to visualize the data using popular libraries such as Matplotlib or Seaborn. Before you choose to visualize your data using a chart, think about what kind of information you are expected to get from the chart and how it can assist you in understanding the data.

As an example, we define three basic charts and their characteristics given in the following subsections.

Box plots

A box plot (`https://www.khanacademy.org/math/statistics-probability/summarizing-quantitative-data/box-whisker-plots/a/box-plot-review`) is an excellent way to visualize and understand data variance. Box plots show results in quartiles, each containing 25% of the values in the dataset; the values are plotted to show how the data is distributed. *Figure 8.2* shows a sample box plot. Note the black dot is an **outlier**:

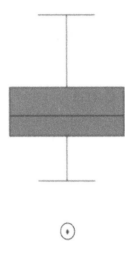

Figure 8.2 – Box plot

The first component of the box plot is the minimum value of the dataset. Then there is the lower quartile, or the minimum 25% values. After that, we have the median value at 50% of the dataset. Then, we have the upper quartile, the maximum 25% value. At the top, we have the maximum value based on the range of the dataset. Finally, we have the outliers. Outliers are the extreme data points—on either the high or low side—that could potentially impact the analysis.

Histograms

A **histogram** represents the numerical data distribution. To create a histogram, you first split the range of values into intervals called **bins**. Once you have defined the number of bins to hold your data, the data is then put into predefined ranges in the appropriate bin. The histogram chart shows the distribution of the data as per the predefined bins. *Figure 8.3* shows a sample histogram. Note that the bins are on the x axis of the plot. The following plot shows the distribution in just two bins. You can see that the distribution is biased toward the first bin.

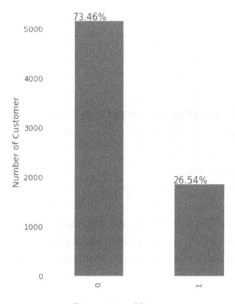

Figure 8.3 – Histogram

Density plots

One of the drawbacks of histograms is that they are sensitive to bin margins and the number of bins. The distribution shape is affected by how the bins are defined. A histogram may be a better fit if your data contains more discrete values (such as gender or postcodes). Otherwise, an alternative is to use a **density plot**, which is a smoother version of a histogram. *Figure 8.4* shows a sample density plot:

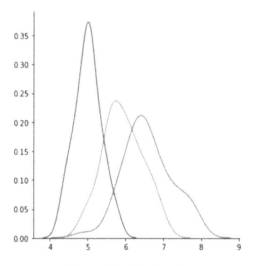

Figure 8.4 – Density plot

Once you have performed the exploratory data analysis, you may choose to go back and collect more data from existing sources or find new sources of data. If you are confident during this stage that the data you have captured can help you achieve the business goal, then you go to the next stage, feature engineering.

Understanding feature engineering

ML is all about data. No matter how advanced our algorithm is, if the data is not correct or not enough, our model will not be able to perform as desired. Feature engineering transforms input data into features that are closely aligned with the model's objectives and converts data into a format that assists in model training.

Sometimes, there is data that may not be useful for a given training problem. How do we make sure that the algorithm is using only the right set of information? What about fields that are not individually useful, but when we apply a function to a group of fields, the data becomes particularly useful?

The act of making your data useful for the algorithm is called feature engineering. Most of the time, a data scientist's job is to find the right set of data for a given problem. Feature engineering requires knowledge of domain-specific techniques, and you will collaborate with business SMEs to better understand the data.

Feature engineering is not only about finding the right features from existing data, but you may need to create new features from existing data. These features are known as **engineered features**.

Imagine that in your flight dataset, there are fields mentioning `scheduled_departure_time` and `departure_time`. Both of these fields will tell you whether the flights are late. However, your business is looking to classify whether the flights are late. You and the business agree to classify the delay into three categories, as follows:

- On time
- Short-delayed
- Long-delayed

A short delay captures the flights that departed with a maximum delay of 30 minutes. All other delayed flights are classified by the long delay value in the delayed column. You will need to add this column or feature to your dataset.

You may end up dropping a column that may not be useful for the given problem. Do you think the `Cancellation Reason` column may be useful for predicting the flight delay? If not, you may choose to drop this column.

You will also represent your data that can be easily digestible by the ML algorithms. A lot of ML algorithms operate on numerical values; however, not all data will be in the numerical format. You will apply techniques such as **one-hot encoding** to convert the columns into a numerical format.

Often, the ML algorithm works well with the value range between -1 and 1 because it is faster to converge and results in better training time. Even if you have numerical data, it could be beneficial to convert it into the range, and the process of doing this is called **scaling**. During this stage, you may write code to scale the dataset.

Data augmentation

In some cases, you may want to create additional records in your datasets for a couple of reasons. One reason is when you do not have enough data to train a meaningful model, while another is when you deliberately want to influence the behavior of the model to favor one answer over the other, such as correcting **overfitting**. This process of creating synthetic data is called **data augmentation**.

All activities related to data collection, processing, cleaning, data analysis, feature engineering, and data augmentation can be done in the platform by using Jupyter notebooks and, potentially, Apache Spark.

Once you have the datasets cleaned, analyzed, and transformed, the next stage is to build and train an ML model.

Building and evaluating the ML model

Congratulations! You are now ready to train your model. You will first evaluate what set of algorithms will be a good fit for the given problem. Is it a regression or classification problem? How do you evaluate to see whether the model is achieving 75% correct predictability as described by the business?

Selecting evaluation criteria

Let's start with accuracy as the model evaluation criteria. This records how many times the predicted values are the same as the labels in the test dataset. However, if the dataset does not have the right variance, the model may guess the majority class for each example, which is effectively not learning anything about the minority class.

You decided to use the confusion matrix to see the accuracy for each class. Let's say you have 1,000 records in your data, out of which 50 are labeled as *delayed*. So, there are 950 examples with the *on time* label. Now, if the model correctly predicts `920` out of 950 for *on time* and `12` out of 50 for the *delayed* label, the matrix will look like the table in *Figure 8.5*:

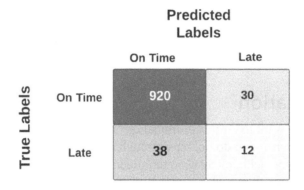

Figure 8.5 – Confusion matrix

For the imbalanced dataset, it is recommended to choose the metrics such as **recall** and **precision** or F-score to get a full picture. In this case, the precision is 31% (12/38) and the recall is 24% (12/50), compared to the accuracy, which is 93.2% (932/1000), and which could be misleading in your scenario.

Building the model

You will start with splitting your data into training, validation, and test sets. Consider a scenario where you split your data into these sets and train a model; let's call this *experiment 1*. Now, you want to retrain the model using different hyperparameters and you split the data again for this new iteration and train the model; let's call it *experiment 2*. Can you compare the results of the two experiments if the data splits across the two experiments are not consistent? It is critical that your data splits are repeatable to compare different runs of your training exercise.

You will try different algorithms or an ensemble of algorithms to assess the performance of the data validation set and review the quality of the predictions. During this stage, every time you try a new adjustment to the model (for example, hyperparameter or different algorithms), you will measure and record the evaluation metrics that were set with the SME during the *Understanding the business problem* stage.

Most of the steps of the modeling stage are iterative. Depending on the result of your experiments, you might realize that the model performance is not as expected. In this case, you may want to go back to the previous steps of the life cycle, such as feature engineering. Or, you may want to redo your data analysis to make sure you understand the data correctly. During training, you will revisit the business objectives and data to find the right balance. You may decide that additional data points from new sources are needed to enhance the training data. It is highly recommended that you present the results to the business stakeholders during this stage. This communication will share the value of the model to the business in the initial stages, collect early feedback, and give the team a chance to course-correct if required.

The next stage is to deploy your model for inferencing.

Deploying the model

Once you have trained your model, the next stage is to version the model in MLflow and deploy it into an environment where the model can be used to make predictions for the incoming requests. The versioning of the models will allow you to keep track of models and roll back to an older version if the need arises.

In this book, we will use the on-line model inference approach. The model has been containerized using the platform's Seldon component and exposed as a REST API. Each call to this REST API will result in one prediction. The stateless container running on Kubernetes will scale hundreds of thousands of requests because of the inherent ability of containers to scale.

The other way is to serve the incoming requests in batches. Imagine a scenario where you have hundreds of thousands of records of labeled data, and you want to test that model behavior for all these records. Making individual REST API calls may not be the right approach in this scenario. Instead, batch inferencing provides an asynchronous approach to making predictions for millions of records. Seldon has the capability to infer batches of data, but it is out of scope for this book.

The REST API you expose for your flight delay prediction could be utilized by the web application to further enhance the customer experience.

Reproducibility

Now, you know what an ML life cycle would look like and how the platform assists you in every step of your journey. As an individual, you may be able to write every step of the data pipelines and model training and tuning in a single notebook. However, this may cause a problem in teams where different people are working on different parts of the life cycle. Let's say someone wants to run the model training part but the entire process is tied up with one another. Your team may not be able to scale with this approach.

A better and more scalable approach is to write different notebooks for various stages (such as data processing and model training) in your project life cycle and use a workflow engine to tie them up. Using the Kubernetes platform, all the stages will be executed using containers and provide a consistent environment for your project between different runs. The platform provides Airflow, an engine that could be used for creating and executing workflows.

Summary

In this short chapter, we wanted to step back and show you the big picture of the platform and the model life cycle. We encourage you to refer to *Chapter 2, Understanding MLOps*, where we presented a typical ML life cycle, for a more detailed discussion. Recall the importance of collaborations across multiple teams and how investing more time in understanding the available data will result in a model that delivers the expected business value.

Now you know what the various stages of your project will look like. In the next two chapters, you will implement the flight delay prediction service using the ML platform that we have presented in this book and you will perform each of the stages we have described in this chapter. The idea is to show you how the platform caters to every stage of your project and how you can implement this platform in your organization.

9
Building Your Data Pipeline

In the previous chapter, you understood the example business goal of improving user experience by recommending flights that have a higher on-time probability. You have worked with the business **subject matter expert** (**SME**) to understand the available data. In this chapter, you will see how the platform assists you in harvesting and processing data from a variety of sources. You will see how on-demand Spark clusters can be created and how workloads could be isolated in a shared environment using the platform. New flights data may be available on a frequent basis and you will see how the platform enables you to automate the execution of your data pipeline.

In this chapter, you will learn about the following topics:

- Automated provisioning of a Spark cluster for development
- Writing a Spark data pipeline
- Using the Spark UI to monitor your jobs
- Building and executing a data pipeline using Airflow

Technical requirements

This chapter includes some hands-on setup and exercises. You will need a running Kubernetes cluster configured with **Operator Lifecycle Manager** (**OLM**). Building such a Kubernetes environment is covered in *Chapter 3*, *Exploring Kubernetes*. Before attempting the technical exercises in this chapter, please make sure that you have a working Kubernetes cluster and **Open Data Hub** (**ODH**) is installed on your Kubernetes cluster. Installing ODH is covered in *Chapter 4*, *The Anatomy of a Machine Learning Platform*.

Automated provisioning of a Spark cluster for development

In this section, you will learn how the platform enables your team to provision an Apache Spark cluster on-demand. This capability of provisioning new Apache Spark clusters on-demand enables your organization to run multiple isolated projects used by multiple teams on a shared Kubernetes cluster without overlapping.

The heart of this component is the Spark operator that is available within the platform. The Spark Kubernetes Operator allows you to start the Spark cluster declaratively. You can find the necessary configuration files in the book's Git repository under the `manifests/radanalyticsio` folder. The details of this operator are out of scope for this book, but we will show you how the mechanism works.

The Spark operator defines a Kubernetes **custom resource definition** (**CRD**), which provides the schema of the requests that you can make to the Spark operator. In this schema, you can define many things, such as the number of worker nodes for your cluster and resources allocated to the master and worker nodes for the cluster.

Through this file, you define the following options. Note that this is not an exhaustive list. For a full list, please look into the documentation of this open source project at `https://github.com/radanalyticsio/spark-operator`:

- The `customImage` section defines the name of the container that provides the Spark software.
- The `master` section defines the number of Spark master instances and the resources allocated to the master Pod.
- The `worker` section defines the number of Spark worker instances and the resources allocated to the worker Pod.
- The `sparkConfiguration` section enables you to add any specific Spark configuration, such as the broadcast join threshold.

- The `env` section enables you to add variables that Spark entertains, such as `SPARK_WORKER_CORES`.

- The `sparkWebUI` section enables flags and instructs the operator to create a Kubernetes Ingress for the Spark UI. In the following section, you will use this UI to investigate your Spark code.

You can find one such file at `manifests/radanalyticsio/spark/cluster/base/simple-cluster.yaml`, and it is shown in the following screenshot. *Figure 9.1* shows a section of the `simple-cluster.yaml` file:

```
manifests > radanalyticsio > spark > cluster > base >  !  simple-cluster.yaml > {} spec > [ ] sparkConfigurat
  1    apiVersion: radanalytics.io/v1
  2    kind: SparkCluster
  3    metadata:
  4      name: spark-cluster-sample
  5      namespace: ml-workshop
  6    spec:
  7      customImage: 'quay.io/ml-on-k8s/spark:3.0.0'
  8
  9      master:
 10        instances: '1'
 11        cpu: '1'
 12        memory: 2Gi
 13      metrics: true
 14      sparkWebUI: false
 15      worker:
 16        instances: '1'
 17        cpu: '1'
 18        memory: 2Gi
 19      env:
 20      - name: SPARK_WORKER_CORES
 21        value: "1"
 22      sparkConfiguration:
 23      - name: spark.sql.conf.autoBroadcastJoinThreshold
 24        value: "20971520"            You, 8 minutes ago • a sample spark cluster file
```

Figure 9.1 – A simple Spark custom resource used by Spark operator

Now, you know the basic process of provisioning a Spark cluster on the platform. However, you will see in the next section that when you select the **Elyra Notebook Image with Spark** notebook image, the Spark cluster is provisioned for you. This is because, in the platform, JupyterHub is configured to submit a Spark cluster **custom resource** (**CR**) when you select a specific notebook. This configuration is available through two files.

The first one is `manifests/jupyterhub/jupyterhub/overlays/spark3/`
`jupyterhub-singleusers-profiles-configmap.yaml`, which defines a profile
as `Spark Notebook`. In this section, the platform configures the name of the container
images under the `images` key, so whenever JupyterHub spawns a new instance of this image,
it will apply these settings. The **Elyra Notebook Image with Spark** notebook points to an
image and it is the same image defined in this part of the configuration. This file contains the
configuration parameters under `configuration`, and the `resources` section points to
resources that will be created alongside the instance of this image. *Figure 9.2* shows a section
of the `jupyterhub-singleusers-profiles-configmap.yaml` file:

```
- name: Spark Notebook
  images:
  - quay.io/ml-aml-workshop/elyra-spark:0.0.4
  env: …


  services:
    spark:
      resources:
      - name: spark-cluster-template
        path: notebookPodServiceTemplate
      - name: spark-cluster-template
        path: sparkClusterTemplate
      configuration:
        worker_nodes: '2'
        master_nodes: '1'
        master_memory_limit: '2Gi'
        master_cpu_limit: '750m'
        master_memory_request: '2Gi'
        master_cpu_request: '100m'
        worker_memory_limit: '2Gi'
        worker_cpu_limit: '2'
        worker_memory_request: '2Gi'
        worker_cpu_request: '1'
        spark_image: 'quay.io/ml-on-k8s/spark:3.0.0'
      return:
        SPARK_CLUSTER: 'metadata.name'
```

Figure 9.2 – A section of jupyterhub-singleusers-profiles-configmap.yaml

Note that `resources` has a property with a value of `sparkClusterTemplate`, which
brings us to our second file.

The second file, `manifests/jupyterhub/jupyterhub/base/jupyterhub-spark-operator-configmap.yaml`, contains `sparkClusterTemplate`, which defines the Spark CR. Note that the parameters available in the `jupyterhub-singleusers-profiles-configmap.yaml` file will be utilized here. *Figure 9.3* shows a section of the `jupyterhub-spark-operator-configmap.yaml` file:

```yaml
sparkClusterTemplate: |
  kind: SparkCluster
  apiVersion: radanalytics.io/v1
  metadata:
    name: "spark-cluster-{{ user }}"
  spec:
    worker:
      instances: "{{ worker_nodes }}"
      memoryLimit: "{{ worker_memory_limit }}"
      cpuLimit: "{{ worker_cpu_limit }}"
      memoryRequest: "{{ worker_memory_request }}"
      cpuRequest: "{{ worker_cpu_request }}"
    master:
      instances: "{{ master_nodes }}"
      memoryLimit: "{{ master_memory_limit }}"
      cpuLimit: "{{ master_cpu_limit }}"
      memoryRequest: "{{ master_memory_request }}"
      cpuRequest: "{{ master_cpu_request }}"
    customImage: "{{ spark_image }}"
    metrics: true
    sparkWebUI: true
    env:
    - name: SPARK_METRICS_ON
      value: prometheus
```

Figure 9.3 – A section of jupyterhub-spark-operator-configmap.yaml

In this section, you have seen how the platform wires different components to make life easier for your teams and organization, and you can change and configure each of these components as per your needs, which brings on the true power of the open source software.

Let's write a data pipeline to process our flights data.

Writing a Spark data pipeline

In this section, you will build a real data pipeline for gathering and processing datasets. The objective of the processing is to format, clean, and transform data into a state that is useable for model training. Before writing our data pipeline, let's first understand the data.

Preparing the environment

In order to perform the following exercises, we first need to set up a couple of things. You need to set up a PostgreSQL database to hold the historical flights data. And you need to upload files to an S3 bucket in MinIO. We used both a relational database and an S3 bucket to better demonstrate how to gather data from disparate data sources.

We have prepared a Postgres database container image that you can run on your Kubernetes cluster. The container image is available at `https://quay.io/repository/ml-on-k8s/flights-data`. It runs a PostgreSQL database with preloaded flights data in a table called `flights`.

Go through the following steps to run this container, verify the database table, and upload CSV files onto MinIO:

1. Run the Postgres database container by running the following command on the same machine where your minikube is running:

    ```
    kubectl create -f chapter9/deployment-pg-flights-data.
    yaml -n ml-workshop
    ```

    You should see a message telling you the `deployment` object is created.

2. Expose the Pods of this deployment through a service by running the following command:

    ```
    kubectl create -f chapter9/service-pg-flights-data.yaml
    -n ml-workshop
    ```

    You should see a message saying that the service object has been created.

3. Explore the contents of the database. You can do this by going inside the Pod, running the Postgres client **command-line interface** (**CLI**), `psql`, and running SQL scripts. Execute the following command to connect to the Postgres Pod and run the Postgres client interface:

    ```
    POD_NAME=$(kubectl get pods -n ml-workshop -l app=pg-
    flights-data)
    ```

4. Connect to the Pod. You can do this by executing the following command:

```
kubectl exec -it $POD_NAME -n ml-workshop -- bash
```

5. Run the Postgres client CLI, `psql`, and verify the tables. Run the following command to log in to the Postgres database from the command line:

```
psql -U postgres
```

This will run the client CLI and connect to the default database.

6. Verify that the tables exist. There should be a table named `flights`. Run the following command from the `psql` shell to verify the correctness of the table:

```
select count(1) from flights;
```

This should give you the number of records in the `flights` table, which is more than 5.8 million, as shown in *Figure 9.4*:

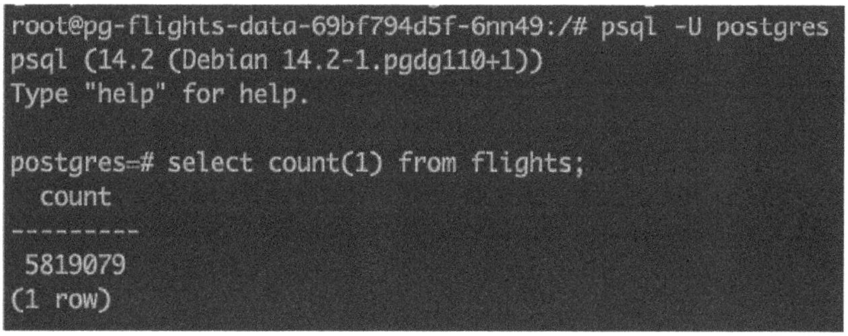

```
root@pg-flights-data-69bf794d5f-6nn49:/# psql -U postgres
psql (14.2 (Debian 14.2-1.pgdg110+1))
Type "help" for help.

postgres=# select count(1) from flights;
  count
---------
 5819079
(1 row)
```

Figure 9.4 – Record count from the flights table

7. Upload the rest of the data to an S3 bucket in MinIO. Open a browser window on the same machine where minikube is running, and navigate to `https://minio.<minikube_ip>.nip.io`. Use the username `minio` and password `minio123`. Remember to replace `<minikube_ip>` with the IP address of your minikube instance.

8. Navigate to **Buckets** and then hit the **Create Bucket** + button. Name the bucket `airport-data` and hit the **Create Bucket** button, as shown in *Figure 9.5*:

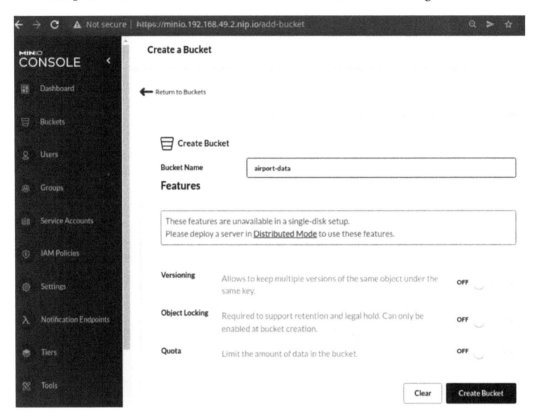

Figure 9.5 – MinIO Create a Bucket dialog

9. While inside the bucket, upload two CSV files from the `chapter9/data/` folder onto the `airport-data` bucket, as shown in *Figure 9.6*:

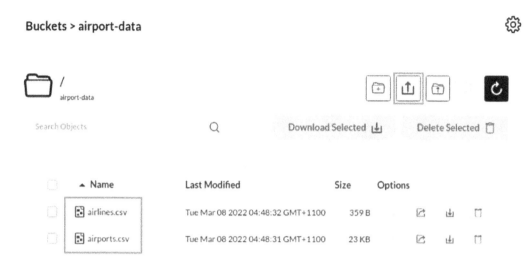

Figure 9.6 – Airport and airline data files

In the real world, you do not need to take the preceding steps. The data sources should already exist and you need to know where to get them. However, for the purpose of the following exercises, we had to load this data into our environment to make it available for the next steps.

You now have the data loaded to the platform. Let's explore and understand the data a little bit more.

Understanding data

Understanding the data includes the following activities. It is important to understand the characteristics of all the datasets involved in order to come up with a strategy and design for the pipeline:

- *Know where the data will be collected from.* Data may come from a variety of sources. It may come from a relational database, object store, NoSQL database, graph database, data stream, S3 bucket, HDFS, filesystem, or FTP. With this information in hand, you will be able to prepare the connectivity you need for your data pipeline. In your case, you need to collect it from a PostgreSQL database and S3 buckets.

- *Understand the format of the data.* Data can come in many shapes and forms. Whether it's a CSV file, a SQL table, a Kafka stream, an MQ stream, a Parquet file, an Avro file, or even an Excel file, you need to have the right tools that can read such a format. Understanding the format helps you prepare the tools or libraries you will need to use to read these datasets.

- *Clean unimportant or irrelevant data.* Understanding what data is important and what is irrelevant helps you design your pipeline in a more efficient way. For example, if you have a dataset with fields for `airline_name` and `airline_id`, you may want to drop `airline_name` in the final output and just use `airline_id` alone. This means one field less to be encoded into numbers, which will improve the performance of model training.

- *Understand the relationships between different datasets.* Identify the identifier fields or primary keys, and understand the join keys and aggregation levels. You need to know this so that you can flatten the data structure and make it easier for the data scientist to consume your datasets.

- *Know where to store the processed data.* You need to know where you will write the processed data so you can prepare the connectivity requirements and understand the interface.

Given the preceding activities, you need a way to access and explore the data sources. The next section will show you how to read a database table from within a Jupyter notebook.

Reading data from a database

Using a Jupyter notebook, let's look at the data. Use the following steps to get started with data exploration, starting with reading data from a PostgreSQL database.

The entire data exploration notebook can be found in this book's Git repository at `chapter9/explore_data.ipynb`. We recommend that you use this notebook to do additional data exploration. It can be by simply displaying the fields, counting the number of occurrences of the same values in a column, and finding the relationships between the data sources:

1. Launch a Jupyter notebook by navigating to `https://jupyterhub.<minikube_ip>.nip.io`. If you are prompted for login credentials, you need to log in with the Keycloak user you've created. The username is `mluser` and the password is `mluser`. Launch the **Elyra Notebook Image with Spark** notebook, as shown in *Figure 9.7*. Because we will be reading a big dataset with 5.8 million records, let's use the **Large** container size. Make sure that, in your environment, you have enough capacity for running a large container. If you do not have enough capacity, try running on a medium container.

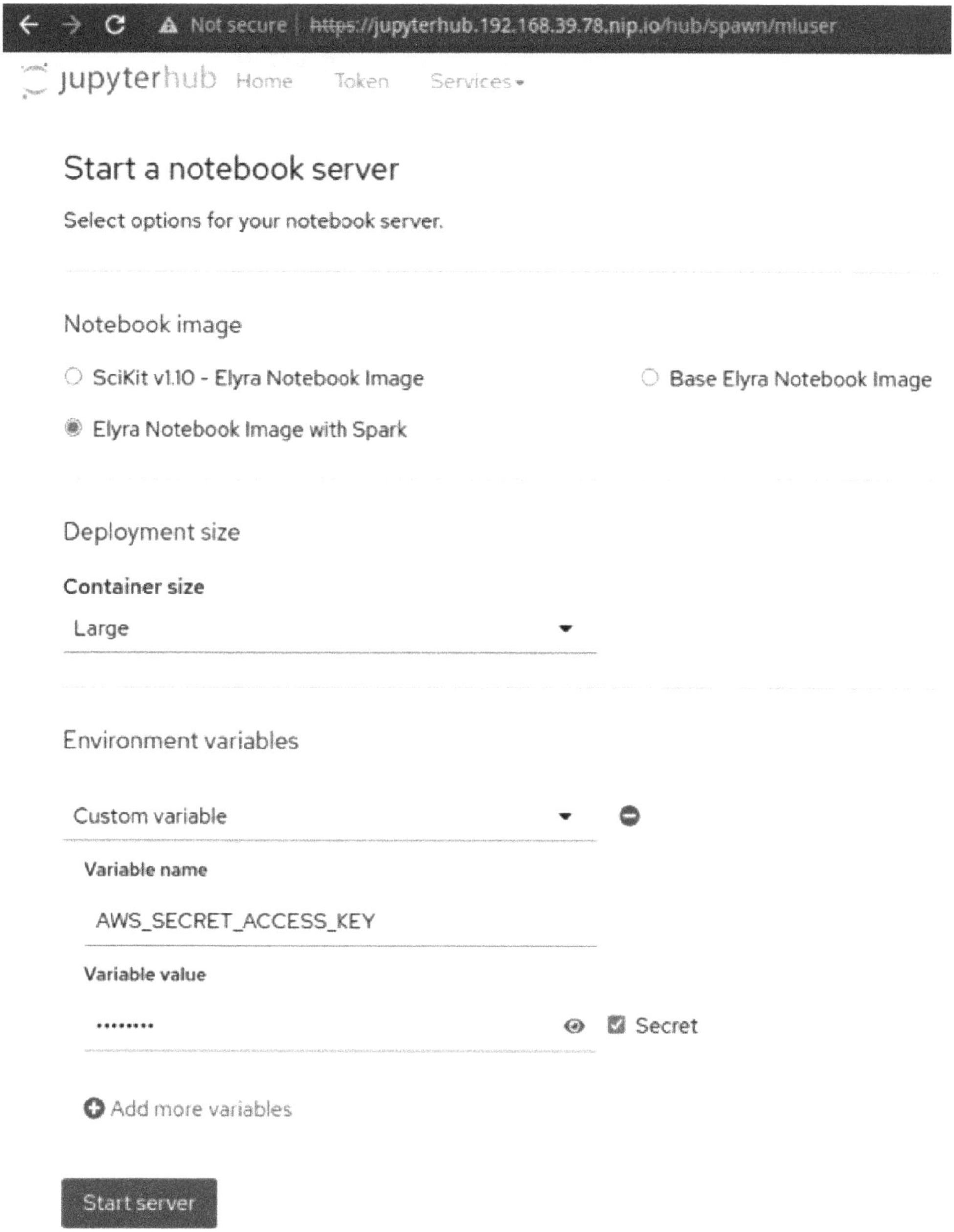

Figure 9.7 – JupyterHub launch page

2. Create a Python 3 notebook. You will use this notebook to explore the data. You can do this by selecting the **File** | **New** | **Notebook** menu option. Then, select **Python 3** as the kernel, as shown in *Figure 9.8*:

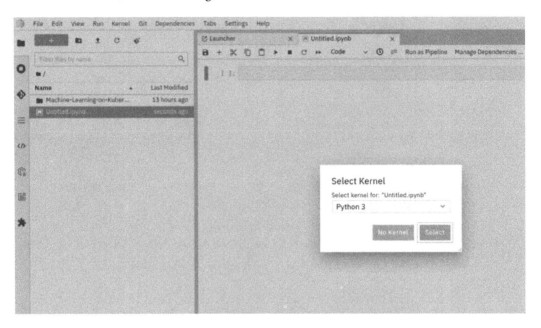

Figure 9.8 – Elyra notebook's kernel selection dialog

3. You can start by looking at the `flights` table in the database. The most basic way of accessing the database is through a PostgreSQL Python client library. Use `psycopg2` for the exercises. You may also choose a different client library to connect to the PostgreSQL database. The code snippet in *Figure 9.9* is the most basic example:

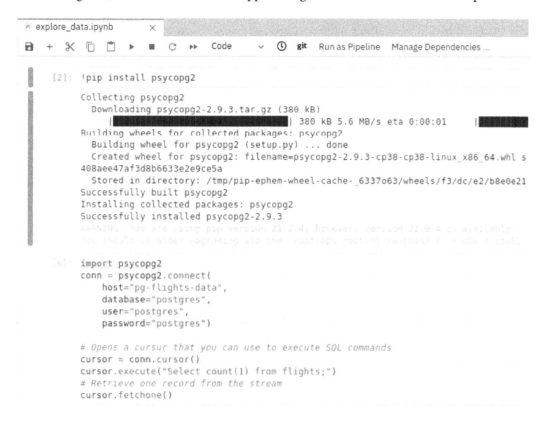

Figure 9.9 – Basic connection to PostgreSQL using psycopg2

4. Another, more elegant, way of accessing the data is through **pandas** or **PySpark**. Both pandas and PySpark allow you to access data, leveraging the functional programming approach through data frames rather than the procedural approach in *Step 3*. The difference between pandas and Spark is that Spark queries can be executed in a distributed manner, using multiple machines or Pods executing your query. This is ideal for huge datasets. However, pandas provides more aesthetically appealing visualizations than Spark, which makes pandas good for exploring smaller datasets. *Figure 9.10* shows a snippet of how to access the database through pandas:

```
[7]: !pip install pandas
Collecting pandas
  Downloading pandas-1.4.1-cp38-cp38-manylinux_2_17_x86_64.manylinux2014_x86_64.whl (11.7 MB)
     |████████████████████████████████| 11.7 MB 2.9 MB/s eta 0:00:01     |███████████|
Requirement already satisfied: pytz>=2020.1 in /opt/app-root/lib/python3.8/site-packages (from panda
Collecting numpy>=1.18.5
  Downloading numpy-1.22.3-cp38-cp38-manylinux_2_17_x86_64.manylinux2014_x86_64.whl (16.8 MB)
     |████████████████████████████████| 16.8 MB 15.4 MB/s eta 0:00:01
Requirement already satisfied: python-dateutil>=2.8.1 in /opt/app-root/lib/python3.8/site-packages (
Requirement already satisfied: six>=1.5 in /opt/app-root/lib/python3.8/site-packages (from python-da
Installing collected packages: numpy, pandas
Successfully installed numpy-1.22.3 pandas-1.4.1
WARNING: You are using pip version 21.2.4; however, version 22.0.4 is available
You should consider upgrading via the '/opt/app-root/bin/python3.8 -m pip install --upgrade pip' com
```

```python
[8]: import psycopg2
import pandas as pds
from sqlalchemy import create_engine

alchemy_engine = create_engine('postgresql+psycopg2://postgres:postgres@pg-flights-data/postgres')
conn = alchemy_engine.connect()
dataFrame = pds.read_sql("select * from flights limit 10;", conn)
print(dataFrame)
conn.close()
```

```
   year  month  day  day_of_week airline  flight_number tail_number  \
0  2015      1    1            4      AS             98      N407AS
1  2015      1    1            4      AA           2336      N3KUAA
2  2015      1    1            4      US            840      N171US
3  2015      1    1            4      AA            258      N3HYAA
4  2015      1    1            4      AS            135      N527AS
5  2015      1    1            4      DL            806      N3730B
6  2015      1    1            4      NK            612      N635NK
7  2015      1    1            4      US           2013      N584UW
8  2015      1    1            4      AA           1112      N3LAAA
9  2015      1    1            4      DL           1173      N826DN

   origin_airport destination_airport scheduled_departure  ... arrival_time  \
0             ANC                 SEA                0005   ...         0408
1             LAX                 PBI                0010   ...         0741
2             SFO                 CLT                0020   ...         0811
3             LAX                 MIA                0020   ...         0756
4             SEA                 ANC                0025   ...         0259
5             SFO                 MSP                0025   ...         0610
6             LAS                 MSP                0025   ...         0509
7             LAX                 CLT                0030   ...         0753
8             SFO                 DFW                0030   ...         0532
9             LAS                 ATL                0030   ...         0656
```

Figure 9.10 – Basic connection to PostgreSQL using pandas

5. If you need to transform a huge dataset, PySpark would be the ideal option for this. For example, let's say you need to transform and aggregate a table with 100 million records. You will need to distribute this work to multiple machines to get faster results. This is where Spark plays an important role. The code snippet in *Figure 9.11* shows how to read the PostgreSQL table through PySpark:

```python
import os
from pyspark.sql import SparkSession

os.environ['PYSPARK_SUBMIT_ARGS'] = "\
--packages org.apache.hadoop:hadoop-aws:3.2.0,org.postgresql:postgresql:42.3.3 \
--master spark://" + os.environ['SPARK_CLUSTER'] + ":7077 pyspark-shell "

# Create the spark application
spark = SparkSession \
    .builder \
    .appName("Python Spark SQL basic example") \
    .getOrCreate()

df = spark.read \
    .format("jdbc") \
    .option("url", "jdbc:postgresql://pg-flights-data:5432/postgres") \
    .option("dbtable", "flights") \
    .option("user", "postgres") \
    .option("password", "postgres") \
    .option("driver", "org.postgresql.Driver") \
    .option("numPartitions", 31) \
    .option("partitionColumn", "day") \
    .option("lowerBound", 0) \
    .option("upperBound", 32) \
    .load()

df.show(5)
df.printSchema()
spark.stop()
```

Figure 9.11 – Reading a PostgreSQL table through PySpark

Because of the distributed architecture of Spark, you need to provide the partitioning information, particularly the number of partitions and the partition column(s), when reading a table from any relational database. Each partition will become a task in Spark's vernacular, and each task can be executed independently by a single CPU core. If the partition information is not provided, Spark will try to treat the entire table as a single partition. You do not want to do this, as this table has 5.8 million records and it may not fit in the memory of a single Spark worker node.

You also need to provide some information about the Spark cluster, such as the master URL and the packages required to run your Spark application. In the example in *Figure 9.12*, we included the `org.postgresql:postgresql:42.3.3` package. This is the PostgreSQL JDBC driver that Spark needs to connect to the database. Spark will automatically download this package from Maven at the application startup.

Reading data from an S3 bucket

Now that you have learned different ways of accessing a PostgreSQL database from a Jupyter notebook, let's explore the rest of the data. While the `flights` table in the database contains the flight information, we also have the *airport* and *airline* information provided as CSV files and hosted in an S3 bucket in MinIO.

Spark can communicate with any S3 server through the `hadoop-aws` library. *Figure 9.12* shows how to access a CSV file in an S3 bucket from a notebook using Spark:

```python
import os
from pyspark.sql import SparkSession


os.environ['PYSPARK_SUBMIT_ARGS'] = f"\
--conf spark.hadoop.fs.s3a.endpoint=http://minio-ml-workshop:9000 \
--conf spark.hadoop.fs.s3a.access.key=minio \
--conf spark.hadoop.fs.s3a.secret.key=minio123 \
--conf spark.hadoop.fs.s3a.path.style.access=true \
--conf spark.hadoop.fs.s3a.impl=org.apache.hadoop.fs.s3a.S3AFileSystem \
--conf spark.hadoop.fs.s3a.multipart.size=104857600 \
--packages org.apache.hadoop:hadoop-aws:3.2.0,org.postgresql:postgresql:42.3.3 \
--master spark://{os.environ['SPARK_CLUSTER']}:7077 pyspark-shell "

# Create the spark application
spark = SparkSession \
    .builder \
    .appName("Python Spark S3 example") \
    .getOrCreate()

dfAirlines = spark.read\
            .options(delimeter=',', inferSchema='True', header='True') \
            .csv("s3a://airport-data/airlines.csv")
dfAirlines.printSchema()

dfAirports = spark.read\
            .options(delimiter=',', inferSchema='True', header='True') \
            .csv("s3a://airport-data/airports.csv")
dfAirports.printSchema()

dfAirports.show(truncate=False)
dfAirlines.show(truncate=False)

print(dfAirports.count())
print(dfAirlines.count())

spark.stop()
```

Figure 9.12 – Spark code to read an S3 bucket from a notebook

Take note that we added a few more Spark submit arguments. This is to tell the Spark engine where the S3 server is and what driver library to use.

After you have explored the datasets, you should have learned the following facts about the data:

- The *flights* table contains 5,819,079 records.

- There are 322 airports in the `airports.csv` file.

- There are 22 airlines in the `airlines.csv` file.

- There is no direct relationship between airports and airlines.

- The `flights` table uses the `IATA_CODE` airport from the `airport` CSV file as the origin and destination airport of a particular flight.

- The `flights` table is using the `IATA_CODE` airline from the `airlines` CSV file to tell which airline is serving a particular flight.

- All the airports are in the United States. This means that the country columns are useless for **machine learning** (**ML**) training.

- The `flights` table has the `SCHEDULED_DEPARTURE`, `DEPARTURE_TIME`, and `DEPARTURE_DELAY` fields, which tell if a flight has been delayed and we can use to produce a `label` column for our ML training.

Given these facts, we can say that we can use both the airports and airline data to add additional airport and airline information to the original `flights` data. This process is usually called **enrichment** and can be done through data frame joins. We can also use the row count information to optimize our Spark code.

Now that you understand the data, you can start designing and building your pipeline.

Designing and building the pipeline

Understanding the data is one thing, designing a pipeline is another. From the data you have explored in the previous section, you learned a few facts. We will use these facts to decide how to build our data pipeline.

The objective is to produce a single, flat dataset containing all the vital information that may be useful for ML training. We said all vital information because we do not know for sure which fields or features are important until we do the actual ML training. As a data engineer, you can take an educated guess, based on your understanding of the data and with the help of an SME, on which fields are important and which ones are not. Along the ML life cycle, the data scientist may get back to you to ask for more fields, drop some fields, or perform some transformation on the data.

With the objective of producing a single dataset in mind, we need to enrich the flight data with the airport and airline data. To enrich the original flight data with airports and airlines data, we need to do a data frame `join` operation. We also need to take note that the flight data has millions of records, while the airport and airline data has less than 50. We can use this information to influence Spark's `join` algorithm for optimization.

Preparing a notebook for Data frame joins

To start, create a new notebook that performs the join, and then adds this notebook as a stage to the pipeline. The following steps will show you how to do this:

1. Create a new notebook. Call it `merge_data.ipynb`.

2. Use Spark to gather the data from the Postgres and S3 buckets. Use the knowledge you learned in the preceding section. *Figure 9.13* shows the data reading part of the notebook. We have also provided a utility Python file, `chapter9/spark_util.py`. This wraps the creation of Spark context to make your notebook more readable. The code snippet in *Figure 9.13* shows you how to use this utility:

```
         ×  │  ▣ merge_data.ipynb      ×

 ⎙  ⎘  ▶  ■  C  ⊮   Code      ∨  ◷  git   Run as Pipeline   Manage Dependencies ...


 import os
 from pyspark.sql import SparkSession
 from pyspark.sql.functions import broadcast
 import spark_util

 submit_args = "--conf spark.hadoop.fs.s3a.endpoint=http://minio-ml-workshop:9000 \
 --conf spark.hadoop.fs.s3a.access.key=minio \
 --conf spark.hadoop.fs.s3a.secret.key=minio123 \
 --conf spark.hadoop.fs.s3a.path.style.access=true \
 --conf spark.hadoop.fs.s3a.impl=org.apache.hadoop.fs.s3a.S3AFileSystem \
 --conf spark.hadoop.fs.s3a.multipart.size=104857600 \
 --packages org.apache.hadoop:hadoop-aws:3.2.0,org.postgresql:postgresql:42.3.3"

 spark = spark_util.getOrCreateSparkSession("Enrich flights data", submit_args)
 spark.sparkContext.setLogLevel("INFO")
 print('Spark context started.')

 Initializing environment variables for Spark
 Creating a spark session...
 Spark session created
```

Spark Context

Cluster Name	spark-cluster-mluser
Version	v3.0.1
Master	spark://spark-cluster-mluser:7077
App Id	app-20220310182619-0005
App Name	Enrich flights data
Driver IP	172.17.0.12

```
 Spark context started.

 df_flights = spark.read \
     .format("jdbc") \
     .option("url", "jdbc:postgresql://pg-flights-data:5432/postgres") \
     .option("dbtable", "flights") \
     .option("user", "postgres") \
     .option("password", "postgres") \
     .option("driver", "org.postgresql.Driver") \
     .option("numPartitions", 31) \
     .option("partitionColumn", "day") \
     .option("lowerBound", 0)\
     .option("upperBound", 31)\
     .load()

 print(f"Partition count:{df_flights.rdd.getNumPartitions()}")

 df_airlines = spark.read\
             .options(delimeter=',', inferSchema='True', header='True') \
             .csv("s3a://airport-data/airlines.csv")
 df_airports = spark.read\
             .options(delimiter=',', inferSchema='True', header='True') \
             .csv("s3a://airport-data/airports.csv")

 #df_flights.printSchema()
 #df_airlines.printSchema()
 #df_airports.printSchema()

 Partition count:31
```

Figure 9.13 – Spark code for preparing the data frames

Notice the new import statement here for broadcast(). You will use this function for optimization in the next step.

3. Perform a data frame join in Spark, as shown in *Figure 9.14*. You need to join all three data frames that you prepared in *Step 2*. From our understanding in the previous section, both the airport and airline data should be merged by `IATA_CODE` as the primary key. But first, let's do the join to the airline data. Notice the resulting schema after the join; there are two additional columns at the bottom when compared to the original schema. These new columns came from the `airlines.csv` file:

```
df_flights = df_flights\
    .join(broadcast(df_airlines), df_flights.airline == df_airlines.IATA_CODE)

df_flights.printSchema()

root
 |-- year: integer (nullable = true)
 |-- month: integer (nullable = true)
 |-- day: integer (nullable = true)
 |-- day_of_week: integer (nullable = true)
 |-- airline: string (nullable = true)
 |-- flight_number: integer (nullable = true)
 |-- tail_number: string (nullable = true)
 |-- origin_airport: string (nullable = true)
 |-- destination_airport: string (nullable = true)
 |-- scheduled_departure: string (nullable = true)
 |-- departure_time: string (nullable = true)
 |-- departure_delay: integer (nullable = true)
 |-- taxi_out: integer (nullable = true)
 |-- wheels_off: string (nullable = true)
 |-- scheduled_time: integer (nullable = true)
 |-- elapsed_time: integer (nullable = true)
 |-- air_time: integer (nullable = true)
 |-- distance: integer (nullable = true)
 |-- wheels_on: string (nullable = true)
 |-- taxi_in: integer (nullable = true)
 |-- scheduled_arrival: string (nullable = true)
 |-- arrival_time: string (nullable = true)
 |-- arrival_delay: integer (nullable = true)
 |-- diverted: integer (nullable = true)
 |-- cancelled: integer (nullable = true)
 |-- cancellation_reason: string (nullable = true)
 |-- air_system_delay: integer (nullable = true)
 |-- security_delay: integer (nullable = true)
 |-- airline_delay: integer (nullable = true)
 |-- late_aircraft_delay: integer (nullable = true)
 |-- weather_delay: integer (nullable = true)
 |-- IATA_CODE: string (nullable = true)
 |-- AIRLINE: string (nullable = true)
```

Figure 9.14 – Spark code for basic data frame join

4. Joining the airport data is a little tricky because you must join it twice: once to
    `origin_airport` and another to `destination_airport`. If we just follow the
    same approach as *Step 3*, the join will work, and the columns will be added to the
    schema. The problem is that it will be difficult to tell which airport fields represent
    the destination airport and which ones are for the airport of origin. *Figure 9.15*
    shows how the field names are duplicated:

Figure 9.15 – Duplicated columns after the join

5. The simplest way to solve this is to create new data frames with prefixed field names (`ORIG_` for origin airports and `DEST_` for destination airports). You can also do the same for the airline fields. *Figure 9.16* shows how to do this:

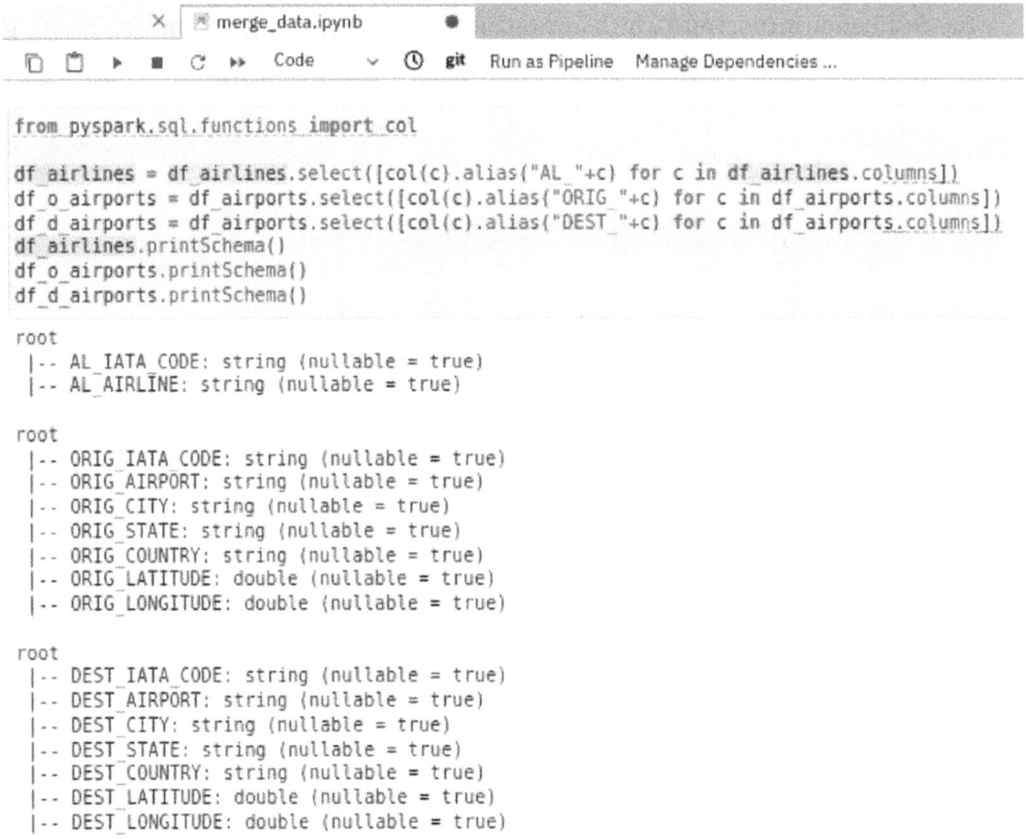
merge_data.ipynb
Code
git
Run as Pipeline
Manage Dependencies ...
from pyspark.sql.functions import col
df_airlines = df_airlines.select([col(c).alias("AL_"+c) for c in df_airlines.columns])
df_o_airports = df_airports.select([col(c).alias("ORIG_"+c) for c in df_airports.columns])
df_d_airports = df_airports.select([col(c).alias("DEST_"+c) for c in df_airports.columns])
df_airlines.printSchema()
df_o_airports.printSchema()
df_d_airports.printSchema()
root
|-- AL_IATA_CODE: string (nullable = true)
|-- AL_AIRLINE: string (nullable = true)
root
|-- ORIG_IATA_CODE: string (nullable = true)
|-- ORIG_AIRPORT: string (nullable = true)
|-- ORIG_CITY: string (nullable = true)
|-- ORIG_STATE: string (nullable = true)
|-- ORIG_COUNTRY: string (nullable = true)
|-- ORIG_LATITUDE: double (nullable = true)
|-- ORIG_LONGITUDE: double (nullable = true)
root
|-- DEST_IATA_CODE: string (nullable = true)
|-- DEST_AIRPORT: string (nullable = true)
|-- DEST_CITY: string (nullable = true)
|-- DEST_STATE: string (nullable = true)
|-- DEST_COUNTRY: string (nullable = true)
|-- DEST_LATITUDE: double (nullable = true)
|-- DEST_LONGITUDE: double (nullable = true)

Figure 9.16 – Adding prefixes to the field names

6. Replace the `df_airports` data frame with `df_o_airports` and `df_d_airports` in your `join` statements, as shown in *Figure 9.17*. Now, you have a more readable data frame:

```
from pyspark.sql.functions import col

df_airlines = df_airlines.select([col(c).alias("AL_"+c) for c in df_airlines.columns])
df_o_airports = df_airports.select([col(c).alias("ORIG_"+c) for c in df_airports.columns])
df_d_airports = df_airports.select([col(c).alias("DEST_"+c) for c in df_airports.columns])
#df_airlines.printSchema()
#df_o_airports.printSchema()
#df_d_airports.printSchema()
```

```
df_flights = df_flights\
    .join(broadcast(df_airlines), df_flights.airline == df_airlines.AL_IATA_CODE)\
    .join(broadcast(df_o_airports), df_flights.origin_airport == df_o_airports.ORIG_IATA_CODE)\
    .join(broadcast(df_d_airports), df_flights.destination_airport == df_d_airports.DEST_IATA_CODE)

df_flights.printSchema()
```

```
root
 |-- year: integer (nullable = true)
 |-- month: integer (nullable = true)
 |-- day: integer (nullable = true)
 |-- day_of_week: integer (nullable = true)
 |-- airline: string (nullable = true)
 |-- flight_number: integer (nullable = true)
 |-- tail_number: string (nullable = true)
 |-- origin_airport: string (nullable = true)
 |-- destination_airport: string (nullable = true)
 |-- scheduled_departure: string (nullable = true)
 |-- departure_time: string (nullable = true)
 |-- departure_delay: integer (nullable = true)
 |-- taxi_out: integer (nullable = true)
 |-- wheels_off: string (nullable = true)
 |-- scheduled_time: integer (nullable = true)
 |-- elapsed_time: integer (nullable = true)
 |-- air_time: integer (nullable = true)
 |-- distance: integer (nullable = true)
 |-- wheels_on: string (nullable = true)
 |-- taxi_in: integer (nullable = true)
 |-- scheduled_arrival: string (nullable = true)
 |-- arrival_time: string (nullable = true)
 |-- arrival_delay: integer (nullable = true)
 |-- diverted: integer (nullable = true)
 |-- cancelled: integer (nullable = true)
 |-- cancellation_reason: string (nullable = true)
 |-- air_system_delay: integer (nullable = true)
 |-- security_delay: integer (nullable = true)
 |-- airline_delay: integer (nullable = true)
 |-- late_aircraft_delay: integer (nullable = true)
 |-- weather_delay: integer (nullable = true)
 |-- AL_IATA_CODE: string (nullable = true)
 |-- AL_AIRLINE: string (nullable = true)
 |-- ORIG_IATA_CODE: string (nullable = true)
 |-- ORIG_AIRPORT: string (nullable = true)
 |-- ORIG_CITY: string (nullable = true)
 |-- ORIG_STATE: string (nullable = true)
 |-- ORIG_COUNTRY: string (nullable = true)
 |-- ORIG_LATITUDE: double (nullable = true)
 |-- ORIG_LONGITUDE: double (nullable = true)
 |-- DEST_IATA_CODE: string (nullable = true)
 |-- DEST_AIRPORT: string (nullable = true)
 |-- DEST_CITY: string (nullable = true)
 |-- DEST_STATE: string (nullable = true)
 |-- DEST_COUNTRY: string (nullable = true)
 |-- DEST_LATITUDE: double (nullable = true)
 |-- DEST_LONGITUDE: double (nullable = true)
```

Figure 9.17 – Updated join statements with prefixed data frames

One thing to note in the join statements is the broadcast() function. In the previous section, we talked about the importance of knowing the sizes of your datasets so that you can optimize your code. The broadcast() function gives a hint to the Spark engine that the given data frame should be broadcasted and that the join operation must use the broadcast join algorithm. This means that before execution, Spark will distribute a copy of the df_airlines, df_o_airports, and df_d_airports data frames to each of the Spark executors so that they can be joined to the records of each partition. In order to make the broadcast join effective, you need to pick the *smaller data frames* to be broadcasted. If you want to know more about this, refer to the performance tuning documentation of Spark in the following URL: https://spark.apache.org/docs/latest/sql-performance-tuning.html.

You have just learned how to join data frames using PySpark. Because PySpark statements are lazily evaluated, the actual execution of the join operations hasn't taken place yet. That is why the printSchema() execution is fast. Spark only performs the processing when the actual data is required. One such scenario is when you persist the actual data to storage.

Persisting the data frames

To get the result of the joins, you need to turn the data frame into physical data. You will write the data frame to S3 storage so that the next stage of your data pipeline can read it. *Figure 9.18* shows a code snippet that writes the joined flights data frame onto a CSV file in MinIO:

```
output_location = "s3a://flights-data/flights"

df_flights.write.mode("overwrite")\
    .option("header","true")\
    .format("parquet").save(output_location)
```

Figure 9.18 – Writing a data frame to an S3 bucket

Executing this will take some time because this is where the actual processing of 5.8 million records takes place. While this is running, you can take a look at what is going on in the Spark cluster. When you started the notebook, it created a Spark cluster in Kubernetes that dedicated the user `mluser` to you. The Spark GUI is exposed at `https://spark-cluster-mluser.<minikube_ip>.nip.io`. Navigate to this URL to monitor the Spark application and to check the status of the application's jobs. You should see one running application named **Enrich flights data**. Clicking on this application name will take you to a more detailed view of the jobs being processed, as shown in *Figure 9.19*:

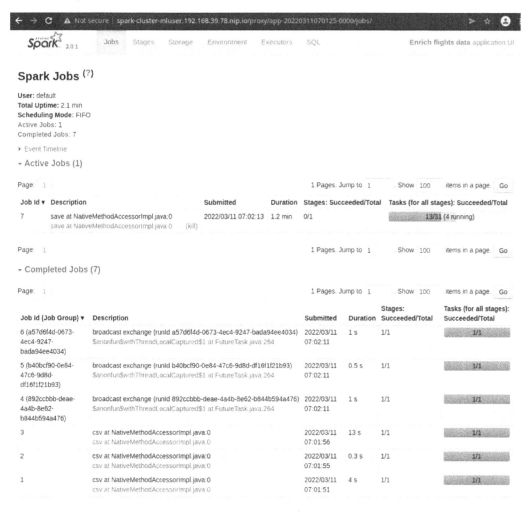

Figure 9.19 – Spark application UI

Figure 9.19 shows the details of the **Enrich flights data** application. Each application is made up of jobs, which are operations. At the bottom of the screen, you can see the **Completed Jobs** section, which includes the broadcast operations. You can also tell that the broadcast operations took around 1 second. Under the **Active Jobs** section, you see the currently running operations, which, in our case, is the actual processing including the reading of the `flights` data from the database, renaming of columns, joining of the data frames, and writing the output to an S3 bucket. This is performed for each partition of the data frame, which translates to **tasks** in Spark. On the right-most column of the **Active Jobs** section, you see the tasks and their progress. Because we partitioned our `flights` data frame by `day of month`, there are 31 partitions. Spark also created 31 parallel processing tasks. Each of these tasks is scheduled to run on **Spark executors**. In *Figure 9.19*, the details say that for the last 1.2 minutes of processing, there are 13 successfully completed tasks out of 31, and there are four currently running.

You may also find tasks that failed in some cases. Failed tasks are automatically rescheduled by Spark to another executor. By default, if the same task fails four times in a row, the whole application will be terminated and marked as failed. There are several reasons task failure happens. Some of them include network interruption or resource congestion, such as out-of-memory exceptions or timeouts. This is why it is important to understand the data so that you can fine-tune the partitioning logic. Here is a basic rule to take note of: the bigger the number of partitions, the smaller the partition size. A smaller partition size will have fewer chances of out-of-memory exceptions, but it also adds more CPU overhead to scheduling. The Spark mechanism is a lot more complex than this, but it is a good start to understanding the relationship between partitions, tasks, jobs, and executors.

Almost half of the data engineering work is actually spent on optimizing data pipelines. There are quite a few techniques to optimize Spark applications, including code optimization, partitioning, and executor sizing. We will not discuss this topic in detail in this book. However, if you want to know more about this topic, you can always refer to the performance tuning documentation of Spark.

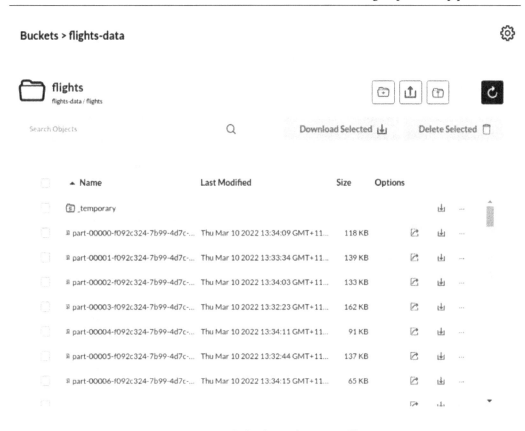

Figure 9.20 – S3 bucket with Parquet files

After the Spark application is completed, the data should be written in S3 in multiple files, with one file representing one partition in Parquet format, as shown in *Figure 9.20*. The **Parquet** file format is a columnar data format, meaning the data is organized by columns rather than by rows as in a typical CSV file. The main advantage of Parquet is that you can cherry-pick columns that you want to read without having to scan the entire dataset. This makes Parquet ideal for analytics, reporting, and also data cleaning, which is what you need to do next.

You can find the full `merge_data.ipynb` notebook in this book's Git repository under the `chapter9` folder. However, we strongly recommend that you create your own notebook from scratch to maximize the learning experience.

Cleaning the datasets

You now have a flat and enriched version of the `flights` dataset. The next step is to clean the data, remove unwanted fields, drop unwanted rows, homogenize the field values, derive new fields, and, perhaps, transform some of the fields.

To start with, create a new notebook and use this notebook to read the Parquet file we generated, and write it as a cleaned version of the dataset. The following steps will walk you through the process:

1. Create a new notebook named `clean_data.ipynb`.

2. Load the `flights` data Parquet files from the `flights-data/flights` S3 bucket, as shown in *Figure 9.21*. Verify the schema and the row count. The row count should be slightly less than the original dataset. This is because the `join` operations performed in the previous steps are inner joins, and there are records in the original `flights` data that do not have airport or airline references.

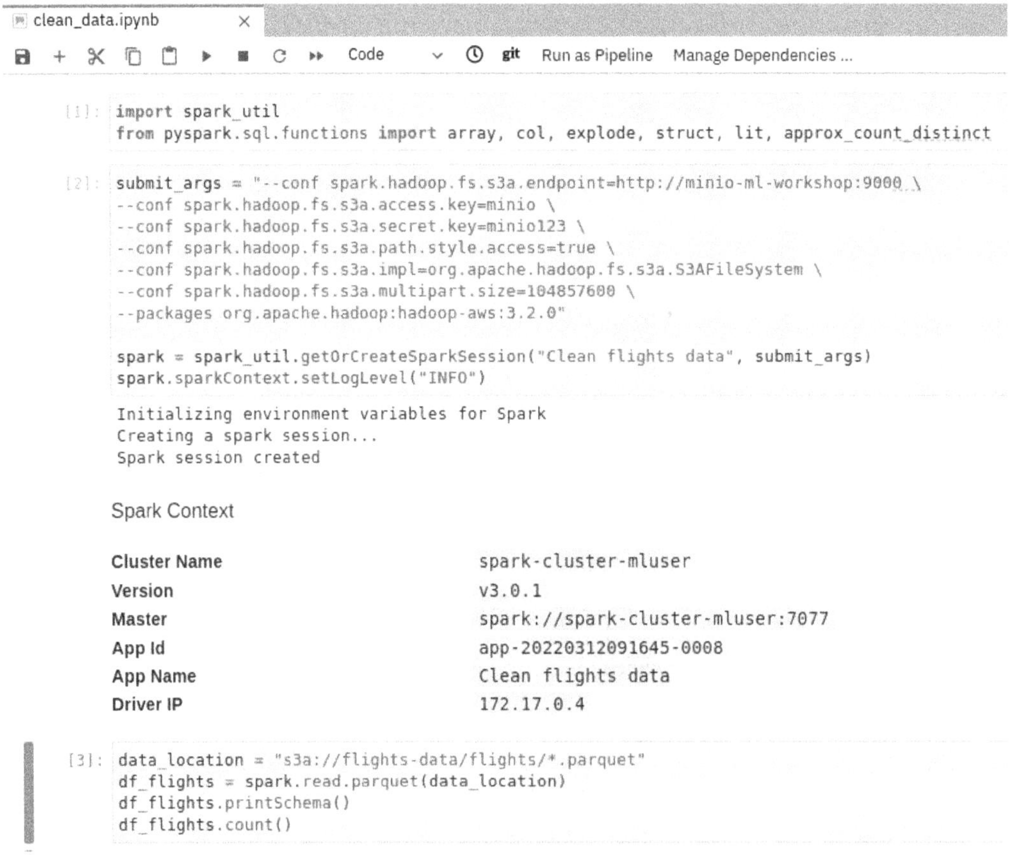

Figure 9.21 – Reading Parquet data from S3

3. Remove the unwanted or duplicated fields, drop fields that have the same value throughout the entire dataset, and create a derived Boolean field called DELAYED, with the value 1 for delayed flights and 0 for non-delayed flights. Let's assume that we only consider a flight as delayed if it is delayed for 15 minutes or more. You can always change this depending on the requirement. Let's do this slowly. Drop the unwanted columns first, as shown in *Figure 9.22*:

```
df_clean = df_flights.drop("AL_IATA_CODE", "ORIG_IATA_CODE", "DEST_IATA_CODE")
df_clean.printSchema()

root
 |-- year: integer (nullable = true)
 |-- month: integer (nullable = true)
 |-- day: integer (nullable = true)
 |-- day_of_week: integer (nullable = true)
 |-- airline: string (nullable = true)
 |-- flight_number: integer (nullable = true)
 |-- tail_number: string (nullable = true)
 |-- origin_airport: string (nullable = true)
 |-- destination_airport: string (nullable = true)
 |-- scheduled_departure: string (nullable = true)
 |-- departure_time: string (nullable = true)
 |-- departure_delay: integer (nullable = true)
 |-- taxi_out: integer (nullable = true)
 |-- wheels_off: string (nullable = true)
 |-- scheduled_time: integer (nullable = true)
 |-- elapsed_time: integer (nullable = true)
 |-- air_time: integer (nullable = true)
 |-- distance: integer (nullable = true)
 |-- wheels_on: string (nullable = true)
 |-- taxi_in: integer (nullable = true)
 |-- scheduled_arrival: string (nullable = true)
 |-- arrival_time: string (nullable = true)
 |-- arrival_delay: integer (nullable = true)
 |-- diverted: integer (nullable = true)
 |-- cancelled: integer (nullable = true)
 |-- cancellation_reason: string (nullable = true)
 |-- air_system_delay: integer (nullable = true)
 |-- security_delay: integer (nullable = true)
 |-- airline_delay: integer (nullable = true)
 |-- late_aircraft_delay: integer (nullable = true)
 |-- weather_delay: integer (nullable = true)
 |-- AL_IATA_CODE: string (nullable = true)
 |-- AL_AIRLINE: string (nullable = true)
 |-- ORIG_AIRPORT: string (nullable = true)
 |-- ORIG_CITY: string (nullable = true)
 |-- ORIG_STATE: string (nullable = true)
 |-- ORIG_COUNTRY: string (nullable = true)
 |-- ORIG_LATITUDE: double (nullable = true)
 |-- ORIG_LONGITUDE: double (nullable = true)
 |-- DEST_AIRPORT: string (nullable = true)
 |-- DEST_CITY: string (nullable = true)
 |-- DEST_STATE: string (nullable = true)
 |-- DEST_COUNTRY: string (nullable = true)
 |-- DEST_LATITUDE: double (nullable = true)
 |-- DEST_LONGITUDE: double (nullable = true)
```

Figure 9.22 – Dropping unwanted columns

We do not need AI_IATA_CODE, ORIG_IATA_CODE, and DEST_IATA_CODE because they are the same as the airline, origin_airport, and destination_airport columns, respectively.

4. Finding the columns with the same values throughout the dataset is an expensive operation. This means you need to count the distinct values of each column for 5 million records. Luckily, Spark provides the approx_count_distinct() function, which is pretty fast. The code snippet in *Figure 9.23* shows how to find the columns with uniform values:

```
df_distinct = df_clean.agg(*(approx_count_distinct(col(c)).alias(c) for c in df_clean.columns))

cols, dtypes = zip(*((c, t) for (c, t) in df_distinct.dtypes))

kvs = explode(array([
    struct(lit(c).alias("column_name"), col(c).alias("distinct_count")) for c in cols
])).alias("kvs")

distinct_count = df_distinct\
    .select([kvs]).select(["kvs.column_name", "kvs.distinct_count"])

uni_value_fields = distinct_count.filter(distinct_count.distinct_count == 1)

uni_value_fields.show(50)

+------------+--------------+
| column_name|distinct_count|
+------------+--------------+
|        year|             1|
|ORIG_COUNTRY|             1|
|DEST_COUNTRY|             1|
+------------+--------------+


col_names = [str(row.column_name) for row in uni_value_fields.select("column_name").collect()]

df_clean = df_clean.drop(*col_names)
df_clean.printSchema()
```

Figure 9.23 – Dropping columns that have uniform values in all rows

5. Finally, create the `label` field that determines whether the flight is delayed or not. The data scientist may use this field as the label for training. However, the data scientist may also use an analog range, such as `departure_delay`, depending on the algorithm chosen. So, let's keep the `departure_delay` field together with the new Boolean field based on the 15-minute threshold on `departure_delay`. Let's call this new field `DELAYED`:

```
delay_threshold = 15


@udf("integer")
def is_delayed(departure_delay, cancelled):
    if(cancelled == 1):
        return 0
    if(departure_delay >= delay_threshold):
        return 1
    return 0


df_clean = df_clean.withColumn("DELAYED", is_delayed(df_clean.departure_delay, df_clean.cancelled))
df_clean.select("month", "day", "flight_number", "departure_delay", "cancelled", "DELAYED").show(50)
```

```
+-----+---+-------------+---------------+---------+-------+
|month|day|flight_number|departure_delay|cancelled|DELAYED|
+-----+---+-------------+---------------+---------+-------+
|    5| 31|         4414|           null|        1|      0|
|    5| 31|         5215|           null|        1|      0|
|    5| 30|          298|             -1|        0|      0|
|    5| 30|         1230|            -11|        0|      0|
|    5| 30|         2044|             17|        0|      1|
|    5| 30|          448|             -6|        0|      0|
|    5| 30|         1126|             -2|        0|      0|
|    5| 30|          612|             31|        0|      1|
|    5| 30|         1747|             -5|        0|      0|
|    5| 30|         2354|             -3|        0|      0|
|    5| 30|         1910|             -5|        0|      0|
|    5| 30|         1279|             -5|        0|      0|
|    5| 30|         1198|             -7|        0|      0|
|    5| 30|          260|              0|        0|      0|
|    5| 30|         2216|             -6|        0|      0|
|    5| 30|          122|             -1|        0|      0|
|    5| 30|          998|             -5|        0|      0|
|    5| 30|          546|             -2|        0|      0|
|    5| 30|          806|             -1|        0|      0|
|    5| 30|         2196|             -3|        0|      0|
|    5| 30|          736|            -10|        0|      0|
|    5| 30|         1254|             -3|        0|      0|
|    5| 30|           20|             54|        0|      1|
|    5| 30|         1965|            -14|        0|      0|
|    5| 30|         2440|             -2|        0|      0|
```

Figure 9.24 – Creating the DELAYED column

Figure 9.24 shows the code snippet for creating a derived column. Test the column creation logic by running a simple query using the `show()` function.

6. Now, write the physical data to the same S3 bucket under the `flights-clean` path. We also want to write the output in Parquet (see *Figure 9.25*):

```
output_location = "s3a://flights-data/flights-clean"
df_clean.cache() #this is to make sure the DAG is not recalculated when we call the .count() later
df_clean.write.mode("overwrite")\
    .option("header","true")\
    .format("parquet").save(output_location)

df_clean.count()
```

5332914

Figure 9.25 – Writing the final data frame to S3

As a data engineer, you need to agree with the data scientist on the output format. Some data scientists may want to get a single huge CSV file dataset instead of multiple Parquet files. In our case, let's assume that the data scientist prefers to read multiple Parquet files.

7. *Step 6* may take quite some time. You can visit the Spark UI to monitor the application execution.

You can find the full `clean_data.ipynb` notebook in this book's Git repository under the `chapter9` folder. However, we strongly recommend that you create your own notebook from scratch to maximize the learning experience.

Using the Spark UI to monitor your data pipeline

While running Spark applications, you may want to look deeper into what Spark is actually doing in order to optimize your pipeline. The Spark UI provides very useful information. The landing page from the master displays the list of worker nodes and applications, as shown in *Figure 9.26*:

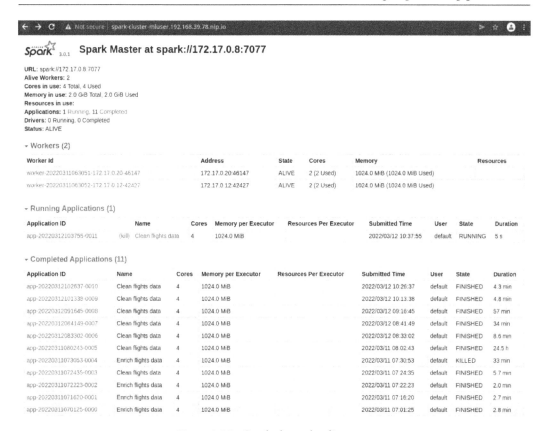

Figure 9.26 – Spark cluster landing page

The landing page also displays the historical application runs. You can see some of the details of the completed application by clicking on one of the completed application IDs. However, we are more interested in the running application when monitoring applications. Let's understand the information in the UI a little bit more.

Exploring the workers page

Workers are machines that are part of the Spark cluster. Their main responsibility is to run executors. In our case, the **worker nodes** are Kubernetes Pods with a worker **Java virtual machine (JVM)** running in them. Each Worker can host one or more executors. However, this is not a good idea when running Spark workers on Kubernetes, so you should configure your executors in a way that only one executor can run in a worker:

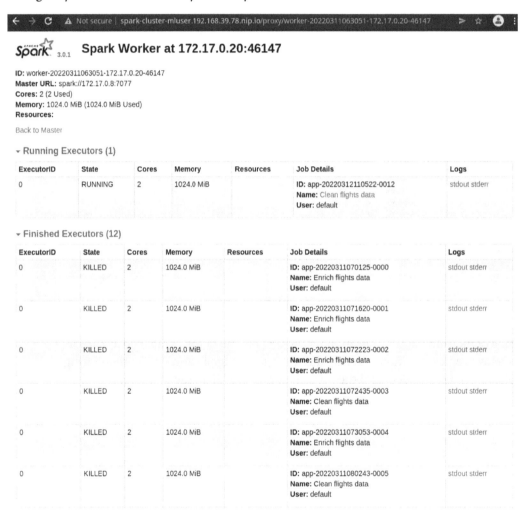

Figure 9.27 – Spark Worker view

Clicking on one of the workers in the UI will take you to the worker UI where you can see all the executors that this worker has run or is currently running. You can also see which application owns the executor. You can see how much CPU or memory is allocated to it, and you can even see the logs of each executor.

Exploring the Executors page

Executors are processes that run inside the worker nodes. Their main responsibility is to execute tasks. An executor is nothing but a Java or JVM process running on the worker node. The worker JVM process manages instances of executors within the same host. Going to `http://spark-cluster-mluser.<minikube_ip>.nip.io/proxy/<application_id>/executors/` will take you to the **Executors** page, which will list all the executors belonging to the current application, as shown in *Figure 9.28*:

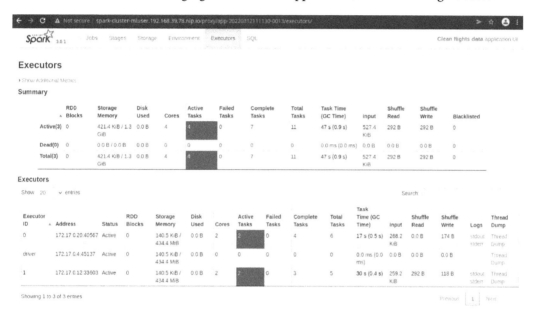

Figure 9.28 – Spark Executors page

On this page, you will find useful metrics that are important in fine-tuning and optimizing your application. For example, you can see the resource usage, garbage collection time, and shuffles. **Shuffles** are exchanges of data across multiple executors, which will happen when you perform an aggregate function, for example. You want to keep this as small as possible.

Exploring the application page

Applications in Spark are any processes that own a Spark context. It could be a running Java, Scala, or Python application that created a Spark session or Spark context and submitted it to the Spark master URL. The applications may not necessarily run in the Spark cluster. It could be anywhere in the network as long as it can connect to the Spark master. However, there is also a mode whereby the application, also called the driver application, is executed inside one of the Spark executors. In our case, the driver application is the Jupyter notebook that is running outside of the Spark cluster. This is why, in *Figure 9.28*, you can see one executor, called **driver**, and not an actual executor ID.

Clicking the application name of a running application from the landing page will bring you to the application UI page. This page displays all the jobs that belong to the current application. A job is an operation that alters the data frame. Each job is composed of one or more tasks. Tasks are a pair of an operation and a partition of a data frame. This is the unit of work that is distributed to the executors. In computer science, this is equivalent to a **closure**. These are shipped over the network as binaries to the worker nodes for the executors to execute. *Figure 9.29* shows the application UI page:

Figure 9.29 – Spark application UI

In the example in *Figure 9.29*, you can see that active job *5* has five tasks, where four tasks are running. The **Tasks** level of parallelism is dependent on the number of CPU cores allocated to the application. You can also get even deeper into a particular job. If you go to `http://spark-cluster-mluser.<minikube_ip>.nip.io/proxy/<application_id>/jobs/job/?id=<job_id>`, you should see the stages of the job and the DAG of each stage.

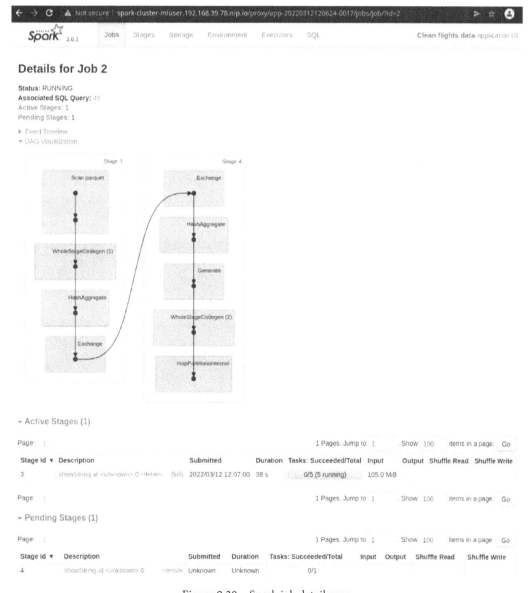

Figure 9.30 – Spark job detail page

The Spark GUI is extremely useful when performing diagnostics and fine-tuning complex data processing applications. Spark is also well documented, and we recommend that you visit Spark's documentation at the following link: `https://spark.apache.org/docs/3.0.0`.

Now that you have created a notebook for enriching the `flights` data and another notebook for cleaning up the dataset to prepare the dataset for the next stage of the ML project life cycle, let's look at how you can automate the execution of these notebooks.

Building and executing a data pipeline using Airflow

In the preceding section, you have built your data pipeline to ingest and process data. Imagine that new `flights` data is available once a week and you need to process the new data repeatedly. One way is to run the data pipeline manually; however, this approach may not scale as the number of data pipelines grows. Data engineers' time would be used more efficiently in writing new pipelines instead of repeatedly running the old ones. The second concern is security. You may have written the data pipeline on sample data and your team may not have access to production data to execute the data pipeline.

Automation provides the solution to both problems. You can schedule your data pipelines to run as required while the data engineer works on more interesting work. Your automated pipeline can connect to production data without any involvement from the development team, which will result in better security.

The ML platform contains Airflow, which can automate the execution and scheduling of your data pipelines. Refer to *Chapter 7*, *Model Deployment and Automation*, for an introduction to Airflow and how the **visual editor** allows the data engineers to build the data pipelines from the same IDE they have used for writing data pipelines. The integration provides the capabilities for data engineering teams to work in a self-serving and independent manner, which further improves the efficiency of your teams.

In the next section, you will automate the data pipeline for the project that you have built in the preceding section.

Understanding the data pipeline DAG

Let's first understand what is involved in running the data pipeline that you have built. Once you have the right information, it would be easy to automate the process.

When you start writing your data pipeline in JupyterHub, you start with the **Elyra Notebook Image with Spark** notebook from the JupyterHub landing page. In the notebook, you connect to the Apache Spark cluster and start writing the data pipelines. The ML platform *knows* that for the **Elyra Notebook Image with Spark** image, it needs to start a new Spark cluster so that it can be used in the notebook. Once you have finished your work, you shut down your Jupyter environment, which results in shutting down the Apache Spark cluster by the ML platform.

The following are three major stages involved in the execution of your data pipeline for the `flights` data:

1. Start the Spark cluster.
2. Run the data pipeline notebook.
3. Stop the Spark cluster.

Figure 9.31 shows the stages of your DAG:

Figure 9.31 – Airflow DAG for the flights project

Each of these stages will be executed by Airflow as a discrete step. Airflow spins a Kubernetes Pod to run each of these stages while you provide the Pod image required to run each stage. The Pod runs the code defined in the Airflow pipeline for that stage.

Let's see what each stage in our DAG is responsible for.

Starting the Spark cluster

In this stage, a new Spark cluster would be provisioned. This cluster will be dedicated to running one Airflow DAG. The role of automation is to submit the request for a new Spark cluster to Kubernetes as a CR. The Spark operator will then provide the cluster, which can be used for the next step in your DAG.

Once the Airflow engine submits the request to create a Spark cluster, it will move to run the second stage.

Running the data pipeline

In this stage, the set of notebooks (`merge_data` and `clean_data`) that you have written earlier in this chapter will be executed by the Airflow DAG. Recall from *Chapter 7, Model Deployment and Automation*, that Airflow uses different operators to run various stages of your automation pipeline (note that Airflow operators are different from Kubernetes Operators). Airflow provides a notebook operator to run the Jupyter notebooks.

The role of automation is to run your data pipeline notebook using the notebook operator. After the data pipeline has finished executing your code, the Airflow engine will move to the next stage.

Stopping the Spark cluster

At this stage, a Spark cluster would be destroyed. The role of automation is to delete the Spark cluster CR created in the first stage of this DAG. The Spark operator will then terminate the cluster that was used to execute the data pipeline in the previous stage.

Next is to define the container images that will be used by Airflow to execute each of these stages.

Registering container images to execute your DAG

You have just built your automation DAG to run your data pipeline, and each stage of this DAG will be executed by running a separate Pod for each stage:

1. To register the container images, first, open the JupyterHub IDE and click on the **Runtime Images** option on the left menu bar. You will see the following screen:

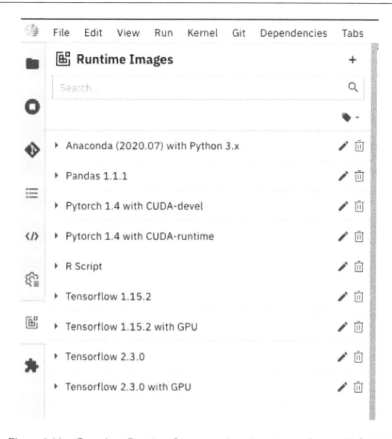

Figure 9.32 – Container Runtime Images registration in your JupyterHub IDE

2. Click on the + icon on the top right to register a new container. You will see the following screen:

Figure 9.33 – Container Runtime Images registration details in your JupyterHub IDE

For the `flights` data pipeline DAG, you will need the following two containers:

I.   The first container image will enable Airflow to run Python code. Fill the screen (shown in *Figure 9.33*) with the following details and click on the button titled **SAVE & CLOSE**:

- **Name**: `AirFlow Python Runner`
- **Description**: `A container with Python runtime`
- **Source**: `quay.io/ml-on-k8s/airflow-python-runner:0.0.11`
- **Image Pull Policy**: **IfNotPresent**

II.   The second container image will enable Airflow to run the data pipeline notebook. Fill the screen shown in *Figure 9.33* with the following details and click on the button titled **SAVE & CLOSE**:

- **Name**: `AirFlow PySpark Runner`
- **Description**: `A container with notebook and pyspark to enable execution of PySpark code`
- **Source**: `quay.io/ml-on-k8s/elyra-spark:0.0.4`
- **Image Pull Policy**: **IfNotPresent**

In the next section, you will build and execute the three stages using Airflow.

Building and running the DAG

In this section, you will build and deploy the DAG using the ML platform. You will first build the DAG using the drag-and-drop editor, and then modify the generated code to further customize the DAG.

Building an Airflow DAG using the visual editor

In this section, you build the DAG for your data processing flow. You will see how JupyterHub assists you in building your DAG using drag-and-drop capabilities:

1.   Start with logging on to JupyterHub on the platform.

2.   Create a new pipeline by selecting the **File | New | PipelineEditor** menu option. You will get a new empty pipeline:

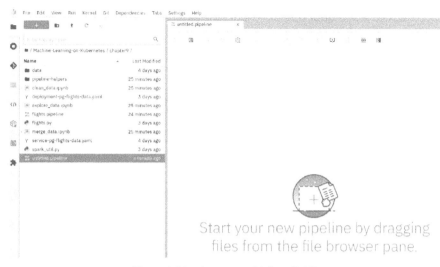

Figure 9.34 – An empty Airflow DAG

3. As shown in the preceding screenshot, you can start by dragging the files required for your pipeline from the file browser on the left-hand side of the editor. For our flights DAG, the first step is to start a new Spark cluster. You will see a file named pipeline-helpers/start-spark-cluster on the browser. Drag it from the browser and drop it on your pipeline:

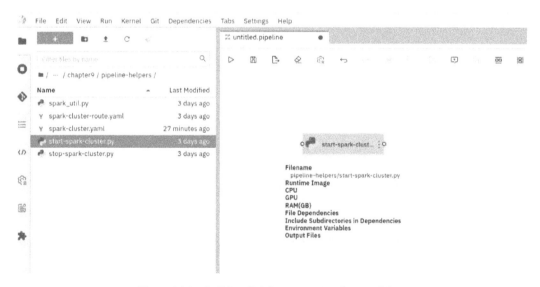

Figure 9.35 – Building DAG stages using drag and drop

4. Complete your pipeline by adding the files that are required for you. The full DAG for the flights data is available in the next step.

5. We have added a pre-built one for you to use as a reference. Go to the folder named Chapter 9/, and open the flights.pipeline file. You can see that there are three stages required for processing the flights data:

Figure 9.36 – DAG view in the JupyterHub IDE

6. Click on the first element of the DAG named **start-spark-cluster**. Right-click on this element and select **Properties**:

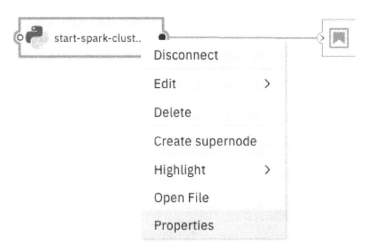

Figure 9.37 – Select the properties of the first stage in your DAG

7.   In the right-hand side window, you can see the properties of this stage:

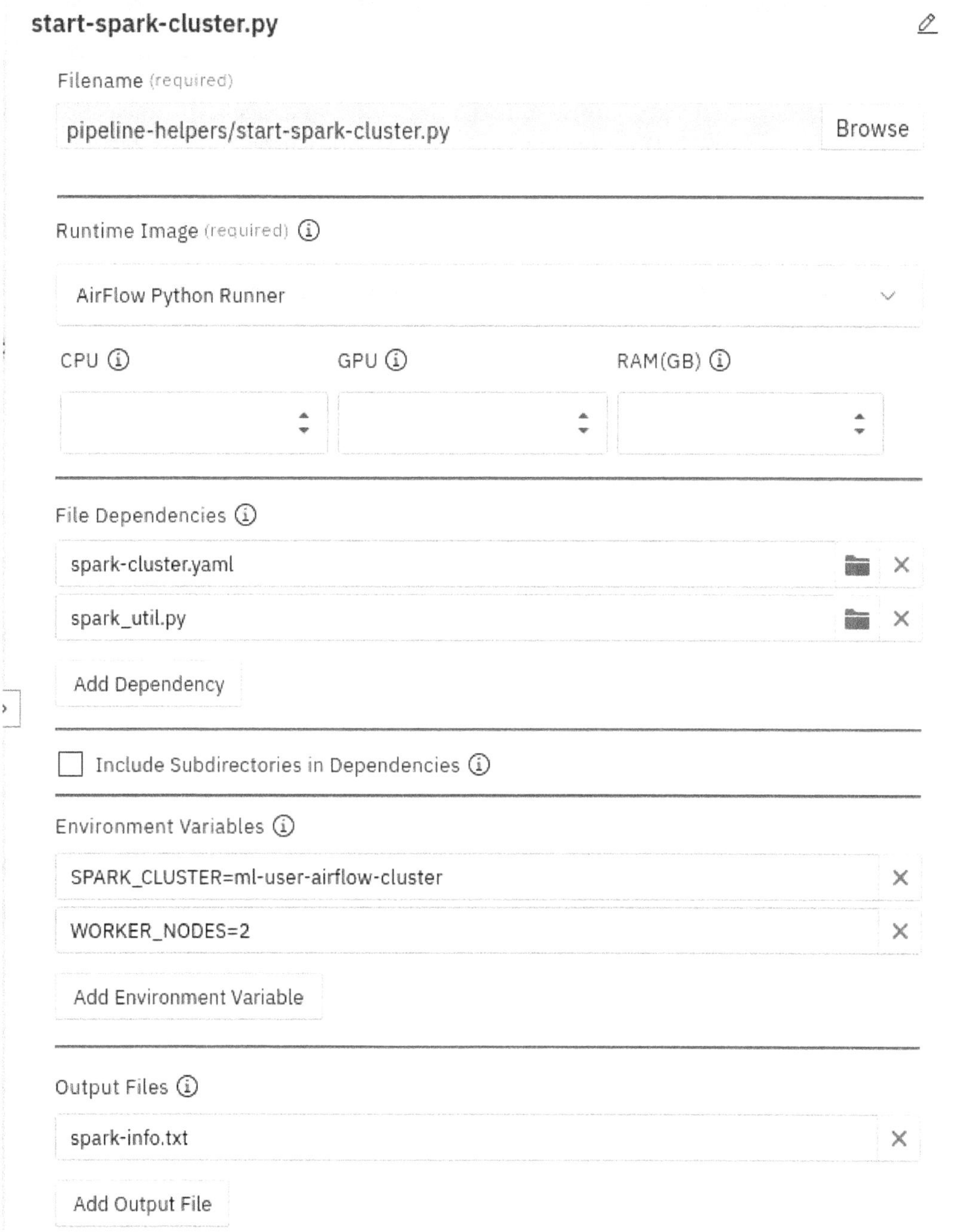

start-spark-cluster.py

Filename (required)

| pipeline-helpers/start-spark-cluster.py | Browse |

Runtime Image (required) ⓘ

AirFlow Python Runner

CPU ⓘ   GPU ⓘ   RAM(GB) ⓘ

File Dependencies ⓘ

spark-cluster.yaml   ✕

spark_util.py   ✕

Add Dependency

☐   Include Subdirectories in Dependencies ⓘ

Environment Variables ⓘ

SPARK_CLUSTER=ml-user-airflow-cluster   ✕

WORKER_NODES=2   ✕

Add Environment Variable

Output Files ⓘ

spark-info.txt   ✕

Add Output File

Figure 9.38 – Properties of the start-spark.py stage

The following list describes each of the properties:

- The **Filename** section defines the file (start-spark-cluster.py) that will be executed by Airflow in this stage.

- The **Runtime Image** section defines the image that will be used to execute the file mentioned in the previous step. This is the container image that you have registered in the earlier section. For the Python stages, you will use the **AirFlow Python Runner** container image.

- The **File Dependencies** section defines files required at this stage. The spark-cluster.yaml defines the configuration of the Spark cluster. The spark_util.py file is the file we have created as a helper utility to talk to the Spark cluster. Note that the files associated with this stage in the DAG will be packaged in the DAG and are available for your stage when it is being executed by Airflow. All of these files are available in the repository.

- The **Environment Variables** section defines environment variables. The file, start-spark-cluster.py in this case, will have access to these environment variables. Think of these variables as configurations that can be used to manage the behavior of your file. For example, the SPARK_CLUSTER variable is used to name the Spark cluster created. WORKER_NODES defines how many worker Pods will be created as Spark workers. So, for bigger jobs, you may choose to change this parameter to have more nodes. Open the start-spark-cluster.py file, and you will see that the two environment variables are being read by it. *Figure 9.39* shows the file:

```
start-spark-cluster.py          ✕

🖫  ▶  ■   Python 3 ∨   Run as Pipeline

 1  import os
 2  import spark_util
 3
 4  cluster_name = os.environ["SPARK_CLUSTER"]
 5  worker_nodes = os.environ["WORKER_NODES"]
 6
 7  if os.getenv("worker_nodes") is None:
 8      worker_nodes = "2"
 9
10  spark_util.start_spark_cluster(cluster_name, worker_nodes)
11  |
```

Figure 9.39 – The start-spark.py file reading the environment variables

The **Output Files** section defines any files created by this stage of the DAG. Airflow will copy this file for all other stages of your DAG. This way you can share the information across multiple stages of your DAG. In this example, the spark_util. py file prints the location of the Spark cluster; think of it as the network name at which the cluster is listening. This name can be used by other stages, such as the data pipeline notebook, to connect to the Spark cluster. There are other options available in Airflow to share data between stages that you can explore and decide the best one for your use case.

8. Click on the second element of the DAG named **merge_data.ipynb**. Right-click on this element and select **Properties**. You will see that for this stage, **Runtime Image** has been changed to **AirFlow PySpark Runner**. You will notice that the file associated with this stage is the Jupyter notebook file. This is the same file you have used to develop the data pipeline. This is the true flexibility of this integration that will take your code as it is to run in any environment.

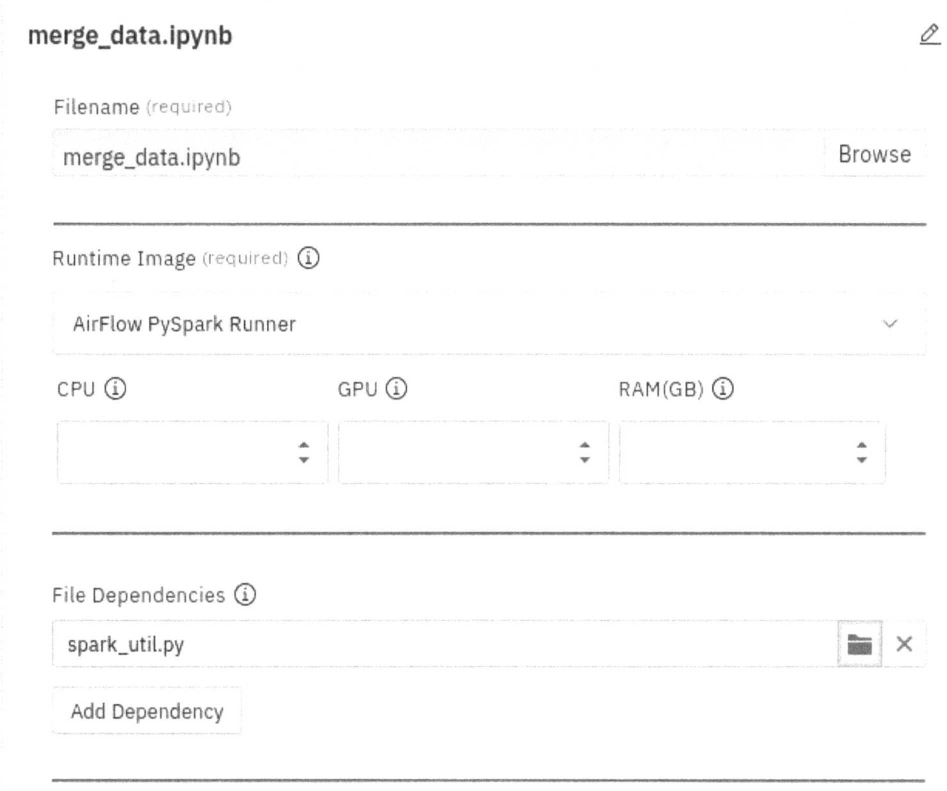

Figure 9.40 – Spark notebook stage in the DAG

Add the second notebook, clean_data.ipynb, as the next stage of the DAG with a similar setup as merge_data.ipynb. We have broken the data pipeline into multiple notebooks for easier maintenance and code management.

9. The last stage of this DAG is stopping the Spark cluster. Notice that **Runtime Image** for this stage is again **AirFlow Python Runner**, as the code is Python-based.

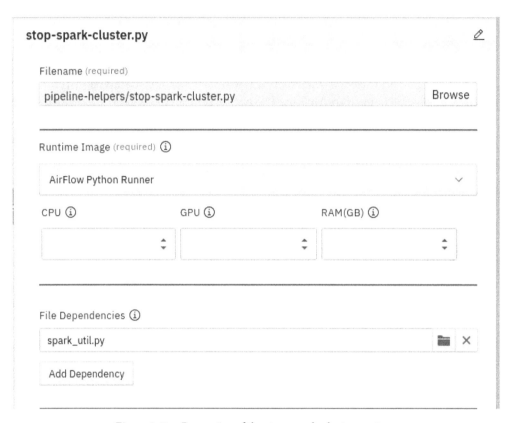

Figure 9.41 – Properties of the stop-spark-cluster.py stage

10. Make sure to save the flights.pipeline file if you make any changes to it.

You have now finished the first DAG. The important thing is that, as a data engineer, you have built the DAG yourself and the data pipeline code you have built is used as it is in the pipeline. This capability will increase the velocity and make your data engineering team autonomous and self-sufficient.

In the next stage, you will run this DAG on the platform.

Running and validating the DAG

In this section, you will run the DAG you have built in the preceding section. We have assumed that you have completed the steps mentioned in *Chapter 7*, *Model Deployment and Automation*, in the *Introducing Airflow* section:

1. Load the `flights.pipeline` file in the JupyterHub IDE and hit the **Run pipeline** icon. The icon is a little *play* button on the icon bar. You will get the following **Run pipeline** screen:

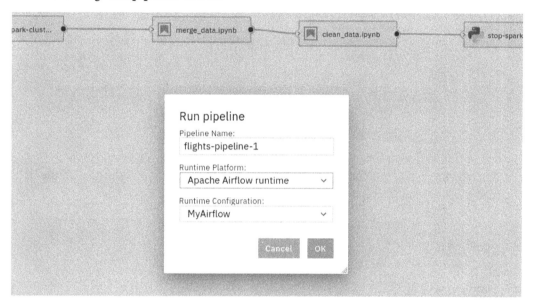

Figure 9.42 – Airflow DAG submission dialog

Give the pipeline a name, select **Apache Airflow runtime** as the **Runtime Platform** option, and select the **Runtime Configuration** option as per your settings. If you have followed the instructions in *Chapter 7*, *Model Deployment and Automation*, then the value would be `MyAirflow`.

2. Click **OK** after you have provided the information.

3. You will see the following screen, validating that the pipeline has been submitted to the Airflow engine in the platform:

Figure 9.43 – Airflow DAG submission confirmation

4. Open the Airflow UI. You can access the UI at `https://airflow.<IP Address>.nip.io`. The IP address is the address of your minikube environment. You will find that the pipeline is displayed in the Airflow GUI:

Figure 9.44 – DAG list in the Airflow GUI

5. Click on the DAG, and then click on the **Graph View** link. You will get the details of the executed DAG. This is the same graph that you have built in the preceding section and has the three stages in it.

Note that your screen may look different depending on your DAG execution stage:

Figure 9.45 – DAG execution status

In this section, you have seen how a data engineer can build the data pipeline (the merge_data notebook) and then is able to package and deploy it using Airflow (flights.pipeline) from the JupyterHub IDE. The platform provides an integrated solution to build, test, and run your data pipelines at scale.

The IDE provides the basics to build the Airflow DAG. What if you want to change the DAG to use the advanced capabilities of the Airflow engine? In the next section, you will see how to change the DAG code generated by the IDE for advanced use cases.

Enhancing the DAG by editing the code

You may have noticed that the DAG that you built ran just once. What if you want to run it on a recurring basis? In this section, you will enhance your DAG by changing its running frequency to run daily:

1. Open flights.pipeline in the JupyterHub IDE. You will see the following familiar screen:

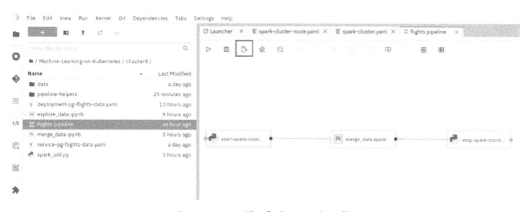

Figure 9.46 – The flights.pipeline file

2. Click on the **Export pipeline** icon on the top bar, and you will be presented with a dialog to export the pipeline. Click on the **OK** button:

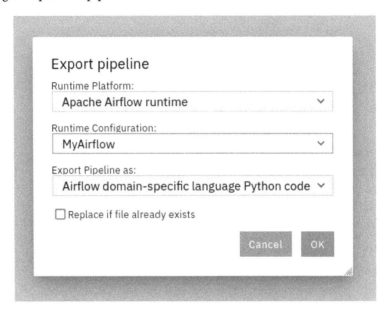

Figure 9.47 – Export pipeline dialog

3. You will get a message that the pipeline export succeeded and a new file will be created as `flights.py`. Open this file by selecting it from the left-hand side panel. You should see the full code of the generated DAG:

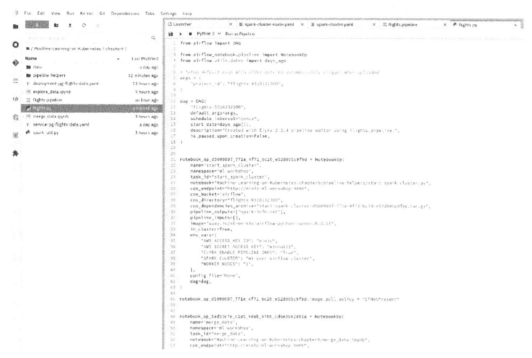

Figure 9.48 – The DAG code after the export

4. You will see your DAG code in Python. From here, you can change the code as needed. For this exercise, we want to change the frequency of the DAG execution. Find the DAG object in the code; it will be around *line 11*:

```
dag = DAG(
        "flights-0310132300",
        default_args=args,
        schedule_interval="@once",
        start_date=days_ago(1),
        description="Created with Elyra 2.2.4 pipeline editor
    using flights.pipeline.",
        is_paused_upon_creation=False,
    )
```

5. Change the schedule of the DAG object. Change the value from `schedule_interval="@once"` to `schedule_interval="@daily"`.

6. The DAG code will look as follows after the change:

```
dag = DAG(
    "flights-0310132300",
    default_args=args,
    schedule_interval="@daily",
    start_date=days_ago(1),
    description="Created with Elyra 2.2.4 pipeline editor
using flights.pipeline.",
    is_paused_upon_creation=False,
)
```

7. Save the file in the IDE and push the file to the Git repository of your DAGs. This is the Git repository that you configured in *Chapter 7, Model Deployment and Automation*, while configuring the Airflow.

8. Now, load the Airflow GUI and you will be able to see your new DAG with the **Schedule** column containing the **@daily** tag. This means that the job will run daily:

Figure 9.49 – Airflow DAG list showing the daily schedule

Congratulations! You have successfully built the data pipeline and automated the execution of the pipeline using the DAG. A big part of this abstraction is the life cycle of the Apache Spark cluster that is managed by the platform. Your team will have a higher velocity because the IDE, automation (Airflow), and data processing engine (Apache Spark) are being managed by the platform.

Summary

Phew! This is another marathon chapter in which you have built the data processing pipeline for predicting flights' on-time performance. You have seen how the platform you have built enables you to write complicated data pipelines using Apache Spark, without worrying about provisioning and maintaining the Spark cluster. In fact, you have completed all the exercises without specific help from the IT group. You have automated the execution of the data pipeline using the technologies provided in the platform and have seen the integration of the Airflow pipelines from your IDE, the same IDE you have used for writing the Spark data pipeline.

Keeping in mind that the main purpose of this book is to help you provide a platform where data and ML teams can work in a self-serving and independent manner, you have just achieved that. You and your team own the full life cycle of data engineering and scheduling the execution of your pipelines.

In the next chapter, you will see how the same principles can be applied to the data science life cycle, and how teams can use this platform to build and automate the data science components for this project.

10
Building, Deploying, and Monitoring Your Model

In the previous chapter, you built the data pipeline and created a basic flight dataset that can be used by your data science team. In this chapter, your data science team will use the flight dataset to build a **machine learning** (**ML**) model. The model will be used to predict the on-time performance of the flights.

In this chapter, you will see how the platform assists you in visualizing and experimenting with the data to build the right model. You will see how to tune hyperparameters and compare the results of different runs of model training. You will see how to register and version models using the components provided by the platform. You will deploy the model as a REST service and start monitoring the deployed model using the components provided by the platform.

Remember that this book is not about data science, instead, the focus is on enabling teams to work autonomously and efficiently. You may see some concepts and steps being repeated from earlier chapters. This is intentional to show you how the concepts provided in the previous chapters help you build a full life cycle.

Keeping the goal in mind, you will learn about the following topics:

- Visualizing and exploring data using JupyterHub
- Building and tuning your model using JupyterHub
- Tracking model experiments and versioning using MLflow
- Deploying your model as a service via Seldon and Airflow
- Monitoring your model using Prometheus and Grafana

Technical requirements

This chapter includes some hands-on setup and exercises. You will need a running Kubernetes cluster configured with the **Operator Lifecycle Manager** (**OLM**). Building such a Kubernetes environment is covered in *Chapter 3, Exploring Kubernetes*. Before attempting the technical exercises in this chapter, please make sure that you have a working Kubernetes cluster and **Open Data Hub** (**ODH**) is installed on your Kubernetes cluster. Installing ODH is covered in *Chapter 4, The Anatomy of a Machine Learning Platform*.

Visualizing and exploring data using JupyterHub

Recall from *Chapter 9, Building Your Data Pipeline*, that the data engineer has worked with the SME of the business and prepared the flight data that can be used to predict the flights' on-time performance.

In this section, you will understand the data produced by the data engineering team. This is the role of the data scientist who is responsible for building the model. You will see how the platform enables your data science and data engineering teams to collaborate and how the data scientist can use the platform to build a model for the given problem.

Let's do some base data exploring using the platform. Keep in mind that the focus of this book is to enable your team to work efficiently. The focus is not on data science or data engineering but on building and using the platform:

1. Launch JupyterHub, but this time select the image that is relative to the data science life cycle. SciKit is one such image available on the platform. Do not click on the **Start server** button just yet.

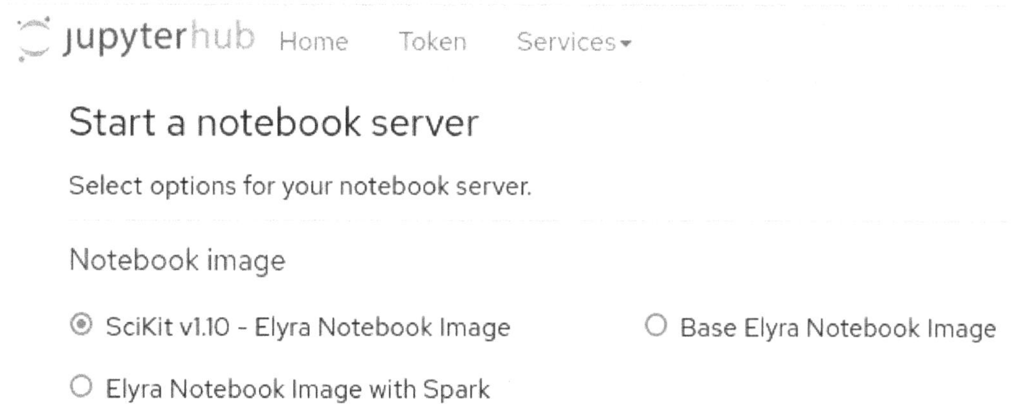

Jupyterhub   Home   Token   Services ▾

Start a notebook server

Select options for your notebook server.

Notebook image

⦿ SciKit v1.10 - Elyra Notebook Image   ○ Base Elyra Notebook Image

○ Elyra Notebook Image with Spark

Figure 10.1 – JupyterHub landing page

2. On the JupyterHub landing page, add an `AWS_SECRET_ACCESS_KEY` variable and populate it with the password for your S3 environment. The value for this key for this exercise would be `minio123`. Notice that we have used the **Medium** container size to accommodate the dataset. Now, hit the **Start server** button to start your JupyterHub IDE.

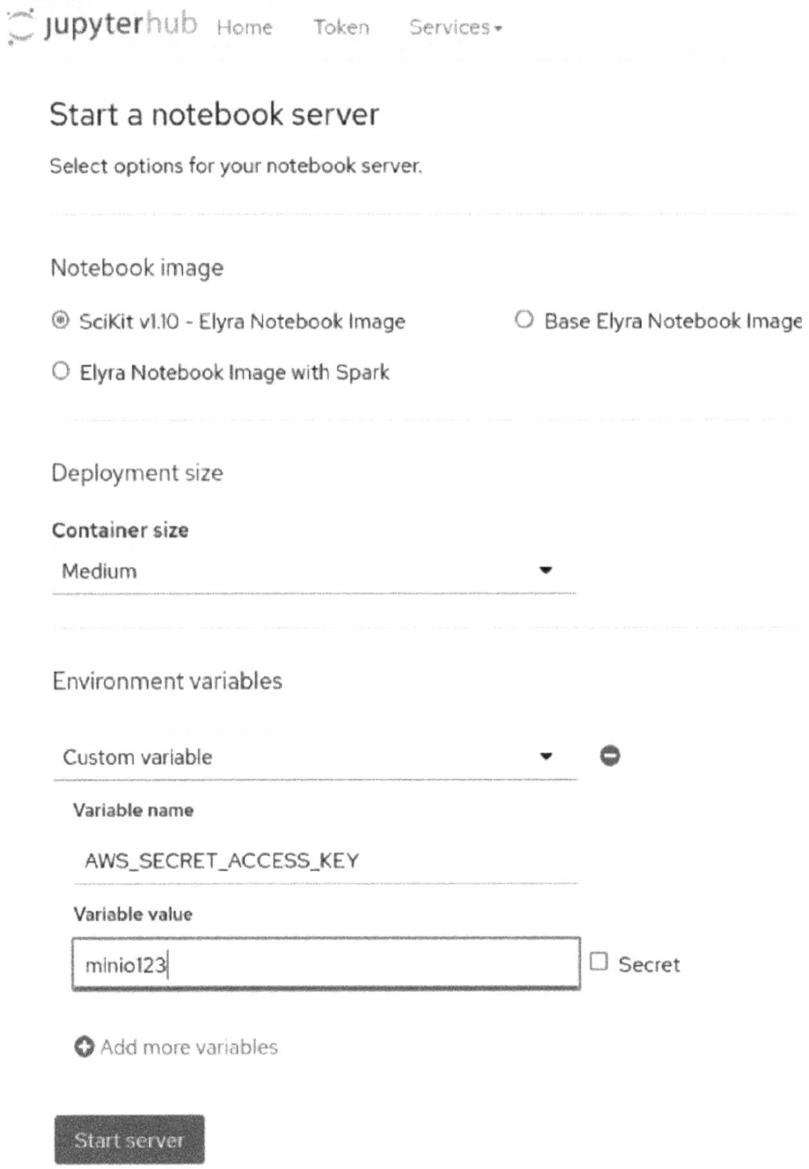

Figure 10.2 – JupyterHub landing page

3. Open the chapter10/visualize.ipynb file notebook in your JupyterHub IDE.

4. The first step is to read the data provided by the data engineering team. Note that the data is available on the same platform, which improves the velocity of the teams. *Cell 2* in the notebook is using the PyArrow library to read the data as a pandas data frame. You will read the data from the flights-data bucket, where data is placed by the data team. You can see the data read code as follows:

```
fs = s3fs.S3FileSystem(anon=True,
                       key='minio',
                       secret='minio123',
                       client_kwargs=dict(endpoint_url="http://minio-ml-workshop:9000"))
df = pq.ParquetDataset('s3://flights-data/flights-clean',
                       filesystem = fs).read_pandas().to_pandas()
```

Figure 10.3 – Cell 2 for the chapter10/visualize notebook

5. The first thing you will do is to look at the data. Trying to make sense of it and familiarizing yourself with what is available can be the ideal take here. You can see in *Cell 3* that the DataFrame's head function has been used to see the first few rows. You will notice the field names and the data in them and see whether you can understand one record. Notice that some fields are NaN and some are None. This gives you a clue that the dataset may not yet be ready for building models. The following screen captures partial output, and it is expected that you run this code in your environment to get the full picture:

```
pd.set_option('display.max_columns', None)
df.head(5)
```

cancelled	cancellation_reason	air_system_delay	security_delay	airline_delay	late_aircraft_delay	weather_delay	AL_AIRLINE	ORIG_AIRPORT	ORIG_CITY
0	None	NaN	NaN	NaN	NaN	NaN	Southwest Airlines Co.	LaGuardia Airport (Marine Air Terminal)	New York
0	None	NaN	NaN	NaN	NaN	NaN	Southwest Airlines Co.	Kansas City International Airport	Kansas City
0	None	NaN	NaN	NaN	NaN	NaN	Southwest Airlines Co.	Orlando International Airport	Orlando
0	None	NaN	NaN	NaN	NaN	NaN	Southwest Airlines Co.	Oakland International Airport	Oakland
0	None	NaN	NaN	NaN	NaN	NaN	Southwest Airlines Co	Will Rogers World Airport	Oklahoma City

Figure 10.4 – Cell 3 for the chapter10/visualize notebook

6. The next stage is to do a simple verification to see how much data is available for you and if you are reading all the records. You can see in *Cell 4* that the DataFrame's `count` function has been used for this. The following screen captures partial output, and it is expected that you run this code in your environment to get the full picture:

```
df.count()

month                  5332914
day                    5332914
day_of_week            5332914
airline                5332914
flight_number          5332914
tail_number            5318547
origin_airport         5332914
destination_airport    5332914
scheduled_departure    5332914
```

Figure 10.5 – Cell 4 for the chapter10/visualize notebook

7. *Cells 5* and *6* are using the DataFrame's shape and the columns' functions are self-explanatory.

8. *Cell 7* is using the DataFrame's `describe` function to generate some basic statistics for the dataset. You may use this to verify whether there is some data that may not make sense. An example could be an exceedingly high value as maximum for the `taxi_in` time. In such cases, you will work with your SME to clarify and adjust the records as needed. The following screen captures partial output, and it is expected that you run this code in your environment to get the full picture:

```
df.describe().T
```

	count	mean	std	min	25%	50%	75%	max
month	5332914.0	6.207210	3.383807	1.00000	3.00000	6.00000	9.00000	12.00000
day	5332914.0	15.688744	8.774687	1.00000	8.00000	16.00000	23.00000	31.00000
day_of_week	5332914.0	3.919179	1.993635	1.00000	2.00000	4.00000	6.00000	7.00000
flight_number	5332914.0	2178.418536	1760.521385	1.00000	731.00000	1690.00000	3252.00000	9855.00000
departure_delay	5249100.0	9.774568	37.592331	-82.00000	-5.00000	-1.00000	8.00000	1988.00000
taxi_out	5246302.0	16.102897	8.991109	1.00000	11.00000	14.00000	19.00000	225.00000
scheduled_time	5332908.0	141.783158	75.251707	18.00000	85.00000	123.00000	174.00000	718.00000

Figure 10.6 – Cell 7 for the chapter10/visualize notebook

9. Next, you want to see whether the data has null values. You have seen in *Step 3*, that there are some NaN and None values in the data. You have found out that there are many columns with missing data problems. The following screen captures partial output, and it is expected that you run this code in your environment to get the full picture:

```
df.isnull().any()

month                   False
day                     False
day_of_week             False
airline                 False
flight_number           False
tail_number             True
origin_airport          False
destination_airport     False
scheduled_departure     False
departure_time          True
departure_delay         True
taxi_out                True
wheels_off              True
```

Figure 10.7 – Cell 8 for the chapter10/visualize notebook

10. You will use the Dataframe's isnull function to find out how many records have this missing data. Using the output from the df.isnull().sum().sort_values(ascending = False) code, there are two different groups. The first six rows of the output show column names that have a very high missing data rate and for these columns, you may talk to data engineering and the SME to find resources from where you can fetch the data for them. For our example, we will just drop these columns.

```
df.isnull().sum().sort_values(ascending = False)

cancellation_reason     5245484
late_aircraft_delay     4329554
weather_delay           4329554
airline_delay           4329554
security_delay          4329554
air_system_delay        4329554
```

Figure 10.8 – Cell 9 for the chapter10/visualize notebook

11. In the second group, starting from the `wheels_on` column, you may either choose to drop the rows containing no data or try to fill the data with a suitable statistics function. For example, the missing `taxi_in` columns could be the mean for the same airport and same time. The strategy must be discussed with the team. For this exercise, we will just drop the rows.

```
wheels_on              89942
arrival_time           89942
taxi_in                89942
wheels_off             86612
taxi_out               86612
departure_time         83814
departure_delay        83814
tail_number            14367
DEST_LATITUDE           4610
DEST_LONGITUDE          4610
ORIG_LATITUDE           4605
ORIG_LONGITUDE          4605
scheduled_time             6
DEST_CITY                  0
DEST_AIRPORT               0
ORIG_STATE                 0
```

Figure 10.9 – Cell 9 for the chapter10/visualize notebook

12. Often, it is a good idea to investigate sample rows where a particular column has no data. You may find a pattern in the data that could be extremely useful in further understanding the data. You have chosen to see the rows where the `tail_number` field has no value and see whether you can find any patterns. The following screen captures partial output, and it is expected that you run this code in your environment to get the full picture:

```
df[df["tail_number"].isna()].head(5)
```

	month	day	day_of_week	airline	flight_number	tail_number	origin_airport	destination_airport	scheduled_departure	departure_time	departure_delay
349	11	30	1	AA	791	None	BOS	PHL	0530	None	NaN
949	11	30	1	UA	600	None	ORD	DCA	0600	None	NaN
1552	11	30	1	UA	263	None	DEN	ORD	1300	None	NaN

Figure 10.10 – Cell 10 for the chapter10/visualize notebook

13. You will then run the Dataframe's `info` function to find out the data types of the columns. A lot of times, the data types of columns are not the ones that you are expecting. You will then talk to the SME and data teams to improve the data quality. The following screen captures partial output, and it is expected that you run this code in your environment to get the full picture:

```
df.info()

<class 'pandas.core.frame.DataFrame'>
RangeIndex: 5332914 entries, 0 to 5332913
Data columns (total 42 columns):
 #   Column               Dtype
---  ------               -----
 0   month                int32
 1   day                  int32
 2   day_of_week          int32
 3   airline              object
 4   flight_number        int32
 5   tail_number          object
 6   origin_airport       object
 7   destination_airport  object
```

Figure 10.11 – Cell 11 for the chapter10/visualize notebook

14. Visualization is one particularly important tool to understand data. You can use any of the libraries that you feel comfortable with. For example, in the last cell of this notebook, you build a graph to find out the data distribution for the DELAYED column. Imagine that 99% of the records are with the DELAYED column as 0. If that is the case, the data may not be enough to predict the flights' on-time performance and you will need to engage the SME and data teams to get more data. For this exercise, we will use the existing data distribution.

```
ax = df["DELAYED"].value_counts().plot(kind='bar',figsize=(14,8),title="Data Distribution")
```

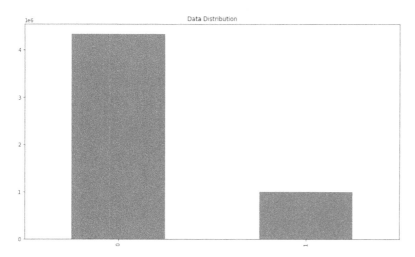

Figure 10.12 – Cell 12 for the chapter10/visualize notebook

Now that we understand flight data a bit better, let's start building our model. In the real world, you would invest a lot more time to understand the data. The focus of this book is to show you how to execute the model development life cycle and so we kept the examples to a minimum.

Building and tuning your model using JupyterHub

As a data scientist, you will want to try different models with different parameters to find the right fit. Before you start building the model, recall from *Chapter 8*, *Building a Complete ML Project Using the Platform*, that you need to define the evaluation criteria, and that **accuracy** may be a misleading criterion for a lot of use cases.

For the flight use case, let's assume that your team and the SME agree on the **PRECISION** metric. Note that precision measures the portion of correct positive identification in the provided dataset.

Let's start writing our model and see how the platform enables data scientists to perform their work efficiently:

1. Open the `chapter10/experiments.ipynb` file notebook in your JupyterHub environment.

2. In *Cell 2*, add the connection information to MLflow. Recall that MLflow is the component in the platform that records the model experiments and works as the model registry. In the code, you will configure EXPERIMENT_NAME, which provides a name for your experiment runs. The last line of this cell mentions how MLflow will record the experiment run. The `autolog` feature enables MLflow to register automatic callbacks during training to record the parameters for later use.

    You also provide the configuration for the S3 bucket, which will be used by MLflow to store the artifacts of your experiments:

```
import mlflow

HOST = "http://mlflow:5500"
EXPERIMENT_NAME = "FlightsDelay-mluser"

os.environ['MLFLOW_S3_ENDPOINT_URL']='http://minio-ml-workshop:9000'
os.environ['AWS_ACCESS_KEY_ID']='minio'
os.environ['AWS_REGION']='us-east-1'
os.environ['AWS_BUCKET_NAME']='mlflow'

# Connect to local MLflow tracking server
mlflow.set_tracking_uri(HOST)

# Set the experiment name...
mlflow.set_experiment(EXPERIMENT_NAME)

mlflow.sklearn.autolog(log_input_examples=True)
```

Figure 10.13 – Cell 2 for the chapter10/experiments notebook

3. *Cell 3* reads the data available from the data engineering team, and *Cell 4* is again providing the information on the missing data from multiple columns. In this notebook, you will use this information to drop the columns that you do not find useful. The following screen captures partial output, and it is expected that you run this code in your environment to get the full picture:

```
import pyarrow.parquet as pq
import s3fs

fs = s3fs.S3FileSystem(anon=True,
                       key='minio',
                       secret='minio123',
                       client_kwargs=dict(endpoint_url="http://minio-ml-workshop:9000"))
fs.ls("flights-data/flights-clean")
#
df = pq.ParquetDataset('s3://flights-data/flights-clean', filesystem = fs).read_pandas().to_pandas()
```

```
df.isnull().sum().sort_values(ascending = False)

cancellation_reason    5245484
late_aircraft_delay    4329554
weather_delay          4329554
airline_delay          4329554
security_delay         4329554
air_system_delay       4329554
air_time                101784
arrival_delay           101784
elapsed time            101784
```

Figure 10.14 – Cell 3 for the chapter10/experiments notebook

4. *Cell 5* is dropping two sets of columns. The first set drops the columns for which you do not have data in most of the rows. You selected these columns based on the previous step. We kept it simple here and dropped the columns; however, it is highly recommended that you work with data teams to find the reason for this anomaly and aim to get as much data as possible. The columns you are dropping are `"cancellation_reason"`, `"late_aircraft_delay"`, `"weather_delay"`, `"airline_delay"`, `"security_delay"`, and `"air_system_delay"`, and are shown in the following screenshot:

```
#drop the columns for which the dataset has missing data for a large number of rows
df = df.drop(["cancellation_reason","late_aircraft_delay","weather_delay",
             "airline_delay", "security_delay", "air_system_delay"], axis=1)
```

Figure 10.15 – Cell 5 for the chapter10/experiments notebook

The second `drop` statement is dropping the `tail_number` column. This column may not play any role in flights getting delayed. In a real-world scenario, you will need to discuss this with the SMEs:

```
# tail number doesnot seems to effect the predictions
df = df.drop(["tail_number"], axis=1)
```

Figure 10.16 – Cell 5 for the chapter10/experiments notebook

5. *Cell 6* is dropping rows for which the data is not available using the Dataframe's `dropna` function. Recall, from *Step 3*, that the number of rows where data is missing from these columns is less compared to the total rows available. `air_time`, `arrival_delay`, and `elapsed_time` are examples of such columns from *Step 5*. We have adopted this approach to keep things simple; a better way would be to find a way to get the missing data or to create this data from existing values.

```
df = df.dropna(subset=["scheduled_time", "ORIG_LONGITUDE", "ORIG_LATITUDE",
                      "DEST_LONGITUDE", "DEST_LATITUDE", "departure_delay",
                      "departure_time", "taxi_out","wheels_off", "taxi_in", "arrival_time",
                      "wheels_on", "elapsed_time", "arrival_delay","air_time"])
```

Figure 10.17 – Cell 6 for the chapter10/experiments notebook

6. In *Cell 7*, you are dropping columns for which you do not have data for future flights. Recall that the model aims to predict the future flight on-time performance. However, columns such as `departure_time` and `arrival_time` contain the actual departure and arrival times. For predicting future flights, you will not have such data available at the time of prediction, and so you need to drop these columns while training your model.

```
df = df.drop(["departure_time","arrival_time","wheels_on",
              "wheels_off", "departure_delay", "arrival_delay", "diverted",
              "cancelled", "taxi_in", "taxi_out"], axis=1)
```

Figure 10.18 – Cell 7 for the chapter10/experiments notebook

7. In the dataset, the scheduled departure and arrival time is available in HHMM format, where HH is hours and MM is minutes. In *Cell 8*, as a data scientist, you may choose to split this data into two different columns where one column represents the hours and the other one represents the minutes. Doing this may simplify the dataset and improve the model performance if some correlation exists between the expected classification and split data. You may do it out of your intuition, or you may discuss this option with the SMEs.

    You have chosen to split the `scheduled_departure` and `scheduled_arrival` columns:

```
df["scheduled_departure_hour"] = df.scheduled_departure.str[:2].astype(int)
df["scheduled_departure_minute"] = df.scheduled_departure.str[2:].astype(int)


df["scheduled_arrival_hour"] = df.scheduled_arrival.str[:2].astype(int)
df["scheduled_arrival_minute"] = df.scheduled_arrival.str[2:].astype(int)
```

Figure 10.19 – Cell 8 for the chapter10/experiments notebook

8. In *Cell 9*, you drop a few more columns. The first set contains columns for which we have to split the time into hours and minutes, such as `scheduled_arrival`:

```
#drop the columns for which we have splitted the time into hours and minutes
df = df.drop(["scheduled_arrival",  "scheduled_departure"], axis=1)
```

Figure 10.20 – Cell 9 for the chapter10/experiments notebook

The second set contains the columns that are represented in other columns. For example, the `origin_airport` column has a key for the airport, and the `ORIG_AIRPORT` column is a descriptive name. Both these columns represent the same information:

```
#drop the columns which are repsented in other column too.
#For example origin_airport col has a key for the airport and the ORIG_A
df = df.drop([ "AL_AIRLINE", "ORIG_AIRPORT", "DEST_AIRPORT"], axis=1)
```

Figure 10.21 – Cell 9 for the chapter10/experiments notebook

9. In *Cell 10*, you visually see the dataset again using the `head` statement. You have noticed that you have some data in string format, such as the `airline` column:

```
pd.set_option('display.max_columns', None)
df.head(5)
```

	month	day	day_of_week	airline	flight_number	origin_airport	destination
0	11	30	1	WN	552	LGA	
1	11	30	1	WN	271	MCI	
2	11	30	1	WN	673	MCO	

Figure 10.22 – Cell 10 for the chapter10/experiments notebook

You choose to encode that data to convert it into numbers. There are many techniques available, such as **ordinal encoding** or **one-hot encoding**, to name a couple. For this example, we choose to use the simple `OrdinalEncoder`. This encoder encodes categorical values as an integer array. In *Cell 12*, you have applied the category encoding to the selected fields such as `airline` and `origin_airport`:

```
import category_encoders as ce

names = ['airline', "origin_airport", "destination_airport", "ORIG_CITY",
        "ORIG_STATE", "DEST_CITY", "DEST_STATE"]
final_df = df
final_df = final_df.drop(['DELAYED'], axis = 1)
enc = ce.ordinal.OrdinalEncoder(cols=names)
enc.fit(final_df)
final_df = enc.transform(final_df)
```

Figure 10.23 – Cell 12 for the chapter10/experiments notebook

This means that the input string data for these fields will be converted into integers. This is good for training; however, at inferencing time, the caller may not know about this encoding that you have just performed. One way is to save this encoder and use it at inferencing time to convert the value from string to integers. So, your inferencing pipeline would consist of two steps. The first step is to apply the encoding and the second step is to predict the response using the saved mode. In the last four lines of *Cell 12*, you have saved the encoder and have to register it with MLflow:

```
#save the encoder to be used at inference time
import joblib
joblib.dump(enc, 'FlgithsDelayOrdinalEncoder.pkl')
#save the file in mlflow
mlflow.log_artifact("FlgithsDelayOrdinalEncoder.pkl")
```

Figure 10.24 – Cell 12 for the chapter10/experiments notebook

10. In *Cell 13*, you validate the data using the `head` statement. Notice that the `airline` column (one of the columns that you have applied the category encoding to) has changed. For example, compare the value of the `airline` column from *Cell 10* and *Cell 13* and notice that the value of **WN** in the `airline` column has been changed to 1. This confirms that the encoding has been applied to the dataset successfully:

```
final_df.head(5)
```

	month	day	day_of_week	airline	flight_number	origin_airport	destination_airport	:
0	11	30	1	1	552	1	1	
1	11	30	1	1	271	2	2	
2	11	30	1	1	673	3	3	
3	11	30	1	1	2720	4	4	
4	11	30	1	1	805	5	5	

Figure 10.25 – Cell 13 for the chapter10/experiments notebook

11. In *Cell 14*, you used the `dftype` statement to validate the data types of each column in the dataset. Many algorithms need data to be in a numerical format and, based on the available models, you may need to move all the fields to a numerical format.

12. In *Cell 15*, you have split your data into training and testing sets. You will train the model using the `X_Train` and `y_train` set and use the `X_Test` and `y_test` for validation of your model performance. You can perform cross-validation to further assess the model performance on unseen data. We assume that you, as a data scientist, are aware of such concepts and, therefore, will not provide more details on this.

```
from sklearn.model_selection import train_test_split

labels = df['DELAYED']
X_train, X_test, y_train, y_test = train_test_split(final_df, labels, test_size=0.2)
```

Figure 10.26 – Cell 15 for the chapter10/experiments notebook

13. In *Cell 16*, you visualize the data distribution of the dataset. The following screenshot captures partial output, and it is expected that you run this code in your environment to get the full picture:

```
ax = X_train["DELAYED"].value_counts().plot(kind='bar',figsize=(14,8),title="D
```

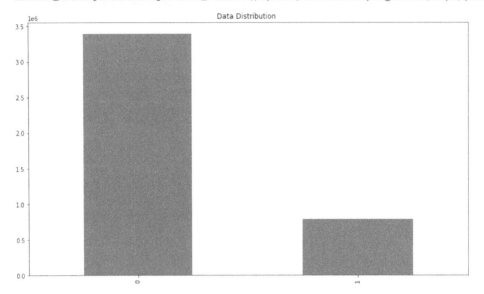

Figure 10.27 – Cell 16 for the chapter10/experiments notebook

You can see from the preceding chart that the data is biased towards the on-time flights. This may impact the performance of the model. Luckily, the RandomForestClassifier object of the SciKit library provides a class_weight parameter. It can take a Python dictionary object where we can provide the desired weights for respective labels. One such example would be to allocate less weight for a value of 0 in the DELAYED column, which represents the on-time flight. A different value for class_weight could be balanced, which will direct the algorithm to weigh the labels as per the inverse proportion to their occurrence frequency. Simply, for our case, the balanced value will put more weight on the value of 1 as compared to the value of 0 in the DELAYED column.

14. In *Cell 19*, you define a random forest classification model and in *Cell 20*, you train the model. You have noticed that we have defined very minimal hyperparameters and then used GridSearchCV to find the best estimator for the given dataset. We have placed a separate set of hyperparameters in the comments of this cell. You are encouraged to try different combinations.

```
from sklearn.ensemble import RandomForestClassifier
from sklearn.model_selection import GridSearchCV

#n_jobs = -1 will use all the cores
forest_clf = RandomForestClassifier(random_state=42, n_jobs=-1, class_weight='balanced')
```

Figure 10.28 – Cell 19 for the chapter10/experiments notebook

Figure 10.29 shows how the model training is performed by executing the `model.fit()` function:

```
# to save time for the example, we have used minimum paramaters
# this surely can be improved for a better model

# criterion = ['gini']
# n_estimators = [15,22]
# max_depth = [15,30]
# min_samples_split = [6,8]
# min_samples_leaf = [10,12]

n_estimators = [15, 22]
criterion = ['gini']
max_depth = [15, 30]

# Merge the list into the variable
hyperparameters = dict(n_estimators = n_estimators , criterion = criterion, max_depth=max_depth)

model = GridSearchCV(forest_clf, hyperparameters, verbose=0)
rf_best_model = model.fit(X_train,y_train)
```

Figure 10.29 – Cell 20 for the chapter10/experiments notebook

The training will take some time to complete, so for *Cell 20*, where you are training your model, be patient.

15. In *Cell 21*, you have used the `predict` method to capture the model prediction for the test data. Note that the `rf_best_model` model is the output of the `GridSearchCV` object:

```
y_test_pred = rf_best_model.predict(X_test)
```

Figure 10.30 – Cell 21 for the chapter10/experiments notebook

16. In *Cell 22*, you have used the `confusion_matrix` function to calculate the matrix and validate the performance of your model:

```
from sklearn.metrics import confusion_matrix, classification_report
confusion_matrix(y_test, y_test_pred)

array([[845005,    1138],
       [195374,    2883]])
```

Figure 10.31 – Cell 22 for the chapter10/experiments notebook

17. In *Cell 23*, you have used the `precision_score` function to calculate `recallscore` for your model on the test dataset. You can see that you have achieved 72% precision, which is good for the first experiment run. You can run more experiments and improve the metrics for your model using the platform:

```
from sklearn.metrics import precision_score, recall_score
precision_score(y_test, y_test_pred)

0.7169858244217856
```

Figure 10.32 – Cell 23 for the chapter10/experiments notebook

You have completed one experiment run with multiple parameters and the `RandomForestClassifier` model. At this stage, you may want to check MLflow and see all the runs the grid search has performed, captured parameters, and model performance data.

Typically, data scientists try multiple algorithms to find the right fit for the given problem. It is up to you to execute and enhance the code and use MLflow to compare different algorithms.

Let's see what MLflow has recorded for us.

Tracking model experiments and versioning using MLflow

In this section, you will use MLflow to track your experiment and version your model. This small section is a review of the capabilities highlighted to you in *Chapter 6*, *Machine Learning Engineering*, where we discussed MLflow in detail.

Tracking model experiments

In this section, you will see the data recorded by MLflow for your experiment. Note that you have just registered the MLflow and called the `autolog` function, and MLflow automatically records all your data. This is a powerful capability in your platform through which you can compare multiple runs and share your findings with your team members.

The following steps shows you how experiment tracking is performed in MLflow:

1. Log in to the MLflow UI of the platform.

2. On the left-hand side, you will see the **Experiments** section and it contains your experiment named **FlightsDelay-mluser**. Click on it and you will see the following screen. The right-hand side shows all the runs. Recall that we have used GridSearchCV so there will be multiple runs:

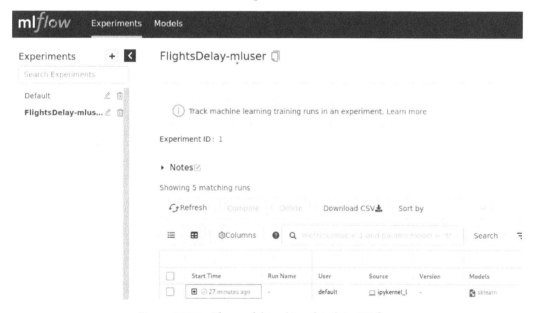

Figure 10.33 – The model tracking details in MLflow

3. Click on the + icon and it will show you all the runs. Based on the hyperparameters, we have four runs, and the best run is automatically selected. As a data scientist, this capability will improve the way you work and provide a system where all the experiments can be recorded, and it is available without too many changes. You have just enabled the `autolog` feature and MLflow will capture the bulk of the metrics automatically. Select all four runs and hit the **Compare** button.

Figure 10.34 shows the comparison of each run and the hyperparameters associated with the run:

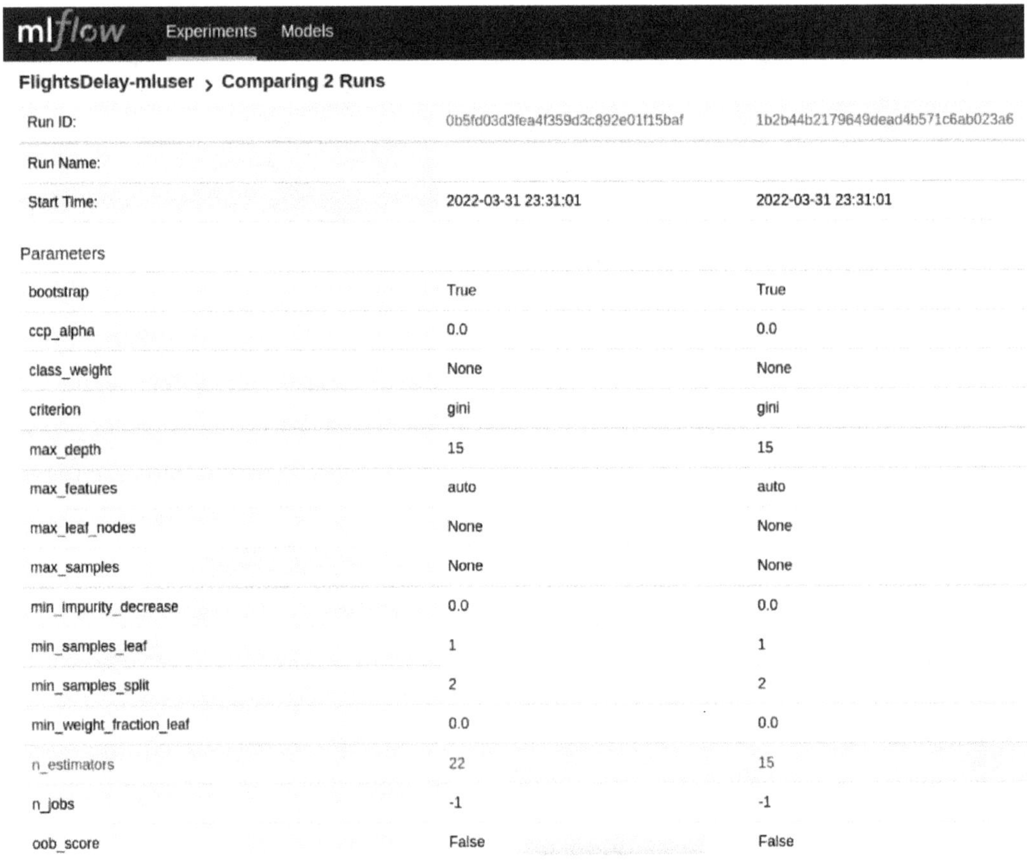

mlflow Experiments Models		
FlightsDelay-mluser ⟩ Comparing 2 Runs		
Run ID:	0b5fd03d3fea4f359d3c892e01f15baf	1b2b44b2179649dead4b571c6ab023a6
Run Name:		
Start Time:	2022-03-31 23:31:01	2022-03-31 23:31:01
Parameters		
bootstrap	True	True
ccp_alpha	0.0	0.0
class_weight	None	None
criterion	gini	gini
max_depth	15	15
max_features	auto	auto
max_leaf_nodes	None	None
max_samples	None	None
min_impurity_decrease	0.0	0.0
min_samples_leaf	1	1
min_samples_split	2	2
min_weight_fraction_leaf	0.0	0.0
n_estimators	22	15
n_jobs	-1	-1
oob_score	False	False

Figure 10.34 – Comparing models in MLflow

4. Click on the run next to the + icon, and MLflow will display the details of this run. In the *artifacts* section, you will find that the model file is available. You can also see that the ordinal encoder file is also available with the name `FlightsDelayOrdinalEncoder.pkl`:

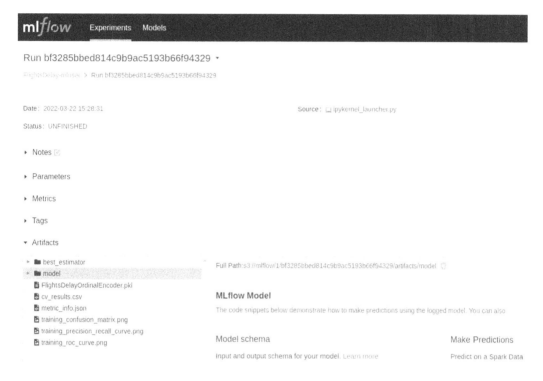

Figure 10.35 – Files and data captured by MLflow

In this section, you have seen that MLflow captured all the metrics from your training run and assisted you in selecting the right model by providing a comparison function.

The next stage is to version your model.

Versioning models

After giving some thought to the model performance and sharing the data with other team members, you have selected the model that can be used for this project. In this section, you will version your model to be used. Refer to *Chapter 6, Machine Learning Engineering*, where we discussed model versioning in detail.

The following steps will guide you on how to version your model:

1. Go to MLflow and click on the **FlightDelay-mluser** experiment on the left-hand side.

2. Then, on the right-hand side of the screen, click on the + icon for your run. You will see the following screen:

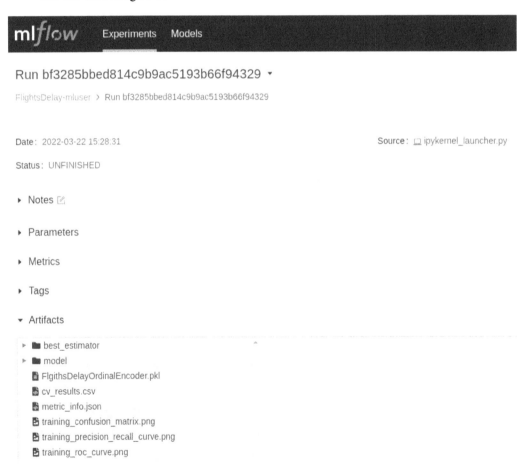

Figure 10.36 – Files and data captured by MLflow

3. Click on the **model** folder under artifacts and a blue button with the **Register Model** label will appear:

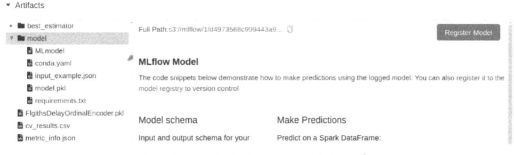

Figure 10.37 – Versioning your models in MLflow

4. Click on the **Register Model** button and enter a name that identifies your model. One example would be `flights-ontime`:

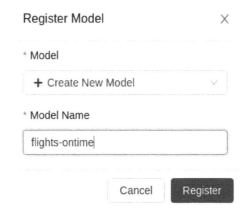

Figure 10.38 – Model registration in MLflow

As a data scientist, you have registered your model for predicting flight delays onto the model registry. The next step is to deploy your model.

Deploying the model as a service

In this section, you will deploy your model as a REST service. You will see that using the details mentioned in *Chapter 7, Model Deployment and Automation*, the team can package and deploy the model as a service. This service will then be consumed by users of your model. We highly encourage you to refresh your knowledge from *Chapter 7, Model Deployment and Automation* before proceeding to this section.

In *Chapter 7, Model Deployment and Automation*, you have deployed the model with a
`Predictor` class, which exposes the model as a REST service. You will use the same class
here, however, in the flight project, you applied categorical encoding to the data before it
was used for model training. This means that you will need to apply the same encoding to
the input data at the inferencing time. Recall that, earlier in this chapter, you saved the file
as `FlightsDelayOrdinalEncoder.pkl` and it is available in the MLflow repository.

The next step is to write a simple class that can apply the transformation to the input data.
Once this class is defined, you will define your inference pipeline using Seldon and then
package your model as a container. So, your inference pipeline will consist of two stages;
the first stage is to apply the encoding and the second stage is to use the model to predict
the class.

Sounds difficult? You will see that the platform abstracts most of the details, and you will
provide a few configuration parameters to package and deploy your model as a service.

Let's first see the `Transformer` class, which will load the
`FlightsDelayOrdinalEncoder.pkl` file and apply the encoding to the input
data. Open the `chapter10/model_deploy_pipeline/model_build_push/`
`Transformer.py` file. You will see that the `__init__` function loads the encoder file
and the `transform_input` function applies the transformation to the input data using
the standard `transform` function. This is the same function you have used during the
model training. *Figure 10.39* shows the code file:

```python
import joblib
import pandas as pd


class Transformer(object):
    """
    this class loads the encoder filesets and apply it to the data passed
    """
    def __init__(self):
        self.encoder = joblib.load('FlightsDelayOrdinalEncoder.pkl')

    def transform_input(self, X, feature_names, meta):
        '''
            Seldon will call this function to apply the transformation
        '''
        df = pd.DataFrame(X, columns=feature_names)
        df = self.encoder.transform(df)
        return df.to_numpy()
```

Figure 10.39 – Transformer class

The second artifact is to define the model inference graph. Recall from *Chapter 7*, *Model Deployment and Automation*, that you have defined a container and one stage in your inference graph using the `SeldonDeploy.yaml` file. In this section, you will extend the inference graph to cater to the transformation and the prediction part of the inference pipeline. Naturally, when you define a new component in your graph, you will also need to define the corresponding container that will be the service for the graph node.

Note that you may choose to execute the transformation logic in `Predict.py` to keep things simple. However, we wanted to show how Seldon can build complicated graphs and each graph could be a separate instance of a container. This approach brings versatility to running your production models in an elastic fashion.

So, let's look into the `chapter10/model_deploy_pipeline/model_deploy/` `SeldonDeploy.yaml` file. This file has been copied from *Chapter 7*, *Model Deployment and Automation*, and the following changes have been made to it.

The first change is to build the inference graph. You need to apply the transformation first and then run the model prediction. *Figure 10.40* displays this graph. Note that the root element for the graph is of the `TRANSFORMER` type with the name `transformer`, and there is a `children` node in the graph. The `children` node will be executed after the root node. This setup allows you to have different graphs as per your model requirements. The child node in this example is the actual prediction:

```yaml
graph:
  name: transformer
  type: TRANSFORMER
  endpoint:
    type: REST
    service_host: localhost
    service_port: 9000
  children:
  - name: predictor
    type: MODEL
    endpoint:
      type: REST
      service_host: localhost
      service_port: 9001
```

Figure 10.40 – Seldon deployment YAML

The second change to the `chapter10/model_deploy_pipeline/model_deploy/SeldonDeploy.yaml` file is registering the containers for both the root and the child node. The `name` field in the graph is the one that associates the container to the graph node. So, we will have two instances of a container, one for `transformer` and the second for `predictor`. The `transformer` instance will execute the `Transformer.py` file and the `predictor` instance will execute the `Predictor.py` file. What we have done is create a single container image with all these files, so our container image is the same. You can examine the `chapter10/model_deploy_pipeline/model_build_push/Dockerfile.py` file where you package all the files into a container image. *Figure 10.41* highlights the part of `SeldonDeploy.yaml` where the containers have been configured.

Note that the first container is with the name `transformer`. The `MODEL_NAME` variable mentions the name of the Python file and the `SERVICE_TYPE` variable mentions the type of callback to call by Seldon. Recall that `Transformer.py` has a `transform_input` method, and `SERVICE_TYPE` guides the Seldon system to call the right function. The same is applied to the `predictor` container instance, and note how `MODEL_NAME` and `SERVICE_TYPE` are different for the `predictor` instance:

```yaml
containers:
  - image: {{ model_coordinates }}
    imagePullPolicy: Always
    name: transformer
    env:
      - name: MODEL_NAME
        value: "Transformer"
      - name: SERVICE_TYPE
        value: TRANSFORMER
      - name: GRPC_PORT
        value: "5007"
      - name: METRICS_PORT
        value: "6007"
      - name: HTTP_PORT
        value: "9000"
  - image: {{ model_coordinates }}
    imagePullPolicy: Always
    name: predictor
    env:
      - name: MODEL_NAME
        value: "Predictor"
      - name: SERVICE_TYPE
        value: MODEL
      - name: GRPC_PORT
        value: "5008"
      - name: METRICS_PORT
        value: "6008"
      - name: HTTP_PORT
        value: "9001"
```

Figure 10.41 – Seldon deployment YAML

That is it! For some of you, this may be a little overwhelming, but once you have defined the structure for your projects, these files can be standardized, and the data scientists will not need to change them for every project. You have seen how the ML platform allows you to be self-sufficient in not only building the models but also packaging them.

The next step is to write a simple Airflow pipeline to deploy your model. Before you start this section, we recommend refreshing your knowledge of deploying the models using Airflow as detailed in *Chapter 7*, *Model Deployment and Automation*. There is no change required in the pipeline that you have built, and you will just be changing a few configuration parameters to provide the right model name and version to the pipeline.

We have prebuilt this pipeline for you, so, open the chapter10/model_deploy_ pipeline/flights_model.pipeline file. Open this file and validate that it has the same two stages as mentioned in *Chapter 7, Model Deployment and Automation*. The first stage builds and pushes the container image to a container registry and the second stage deploys the model using Seldon.

Figure 10.42 displays the first stage with the parameters used for building and pushing the container image. **Runtime Image** and **File Dependencies** have the same values as shown earlier. Notice the **Environment Variables** section, where you have the same variable names but different values:

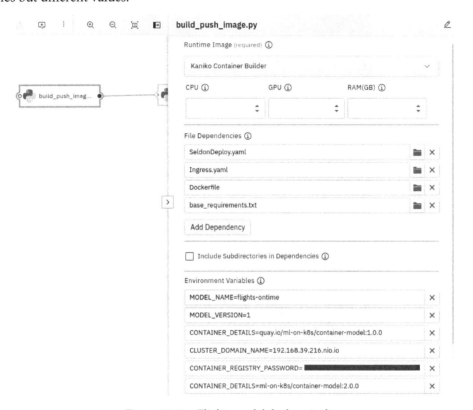

Figure 10.42 – Flights model deploy pipeline

Let's see each of them:

- MODEL_NAME has a value of flights-ontime. This is the name of the model you were given when you registered the model with MLflow.

- MODEL_VERSION has a value of 1. This is the version of the model you would like to deploy. This version is recorded in the MLflow system.

- CONTAINER_DETAILS has a value of flights-ontime. This is the name of the model you were given when you registered the model with MLflow.

- CONTAINER_REGISTRY is the container registry API endpoint. For DockerHub, this is at https://index.docker.io/v1. Set the value of this variable to https://index.docker.io/v1/. In this example, we have used quay.io as the registry. This is another free registry that you can use.

- CONTAINER_REGISTRY_USER is the username of the user that will push images to the image registry. Set this to your DockerHub username or Quay username.

- CONTAINER_REGISTRY_PASSWORD is the password of your container registry user. In production, you do not want to do this. You may use secret management tools to serve your password.

CONTAINER_DETAILS is also the name of the repository to where the image will be pushed and the image name and image tag. *Figure 10.43* displays the second stage with the parameters used for deploying the container image using Seldon. **Runtime Image** and **File Dependencies** have the same values as shown earlier. Notice the **Environment Variable** is the section where you have variable values set for this deployment. The required variables are MODEL_NAME, MODEL_VERSION, CONTAINER_DETAILS, and CLUSTER_DOMAIN. You have seen all the variables in the preceding paragraph, but CLUSTER_DOMAIN is the DNS name of your Kubernetes cluster. In this case, the IP address of minikube is <Minikube IP>.nip.io.

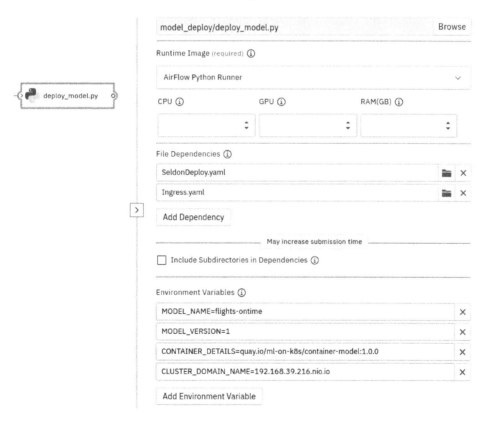

Figure 10.43 – Flights model deploy pipeline

Save and deploy this DAG to your Airflow environment and the model will be available for consumption when the Airflow DAG has finished execution. Validate that this DAG has been executed correctly by logging into Airflow and checking the status of the DAG. *Figure 10.44* shows the Airflow UI where you have validated the DAG's status. Notice that we have saved the DAG under the name `flights-model-deploy`; if you have chosen some other name, your DAG name will reflect accordingly.

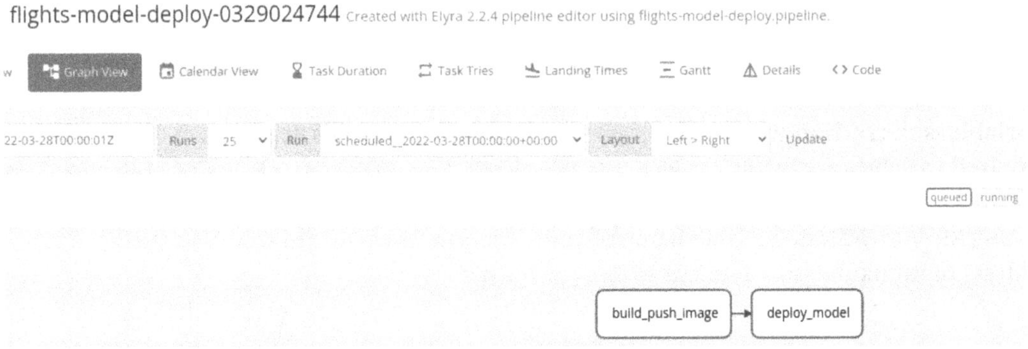

Figure 10.44 – Airflow DAG for the flights pipeline

Recall that MLflow associates a run ID for each of the experiments. You register one of these experiments in the model registry so it can be deployed. Refer to *Figure 10.34*, which shows a screenshot of the run ID for this model.

This model run will be associated with the deployed model, so your team can track the models running in the environment to an individual run. This capability provides a trace back on what version of the model is running in different environments. Run the following command to see the resources created by the model:

```
kubectl get service,ingress,SeldonDeployment -n ml-workshop |
grep bf32
```

You should get the following response. As you can see, the Kubernetes service and ingress have a run ID that starts with `bf32` for this example. Note that it will have a different value for your case, and you will need to adjust the run ID in the preceding command:

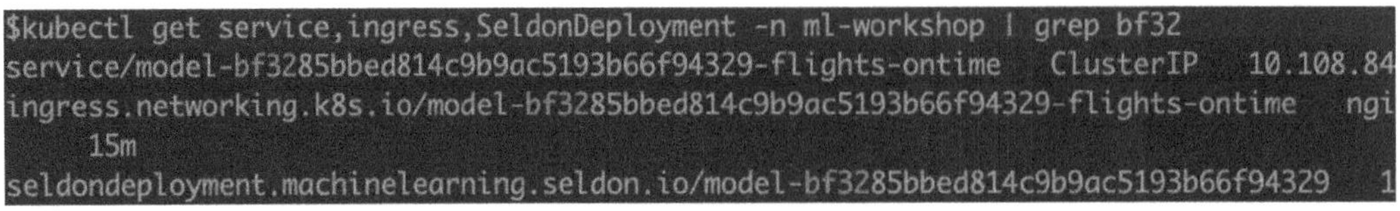

Figure 10.45 – Kubernetes objects created by the platform

Now, the model is deployed; you now test the model by running a RESTful call to your model.

Calling your model

Recall that the model is exposed via the Kubernetes Ingress, which is created by automation. In order to test whether the model is running properly as a RESTful API, follow these steps:

1. Run the following command to get the `ingress` object. Note that the name of the `ingress` object will be different for your setup:

    ```
    kubectl get ingress <INGRESS_NAME> -n ml-workshop
    ```

2. Now, make an HTTP call to the location where your model is available for inference. Run the following commands. The `chapter10/inference` folder contains a payload for the flight data and in return, the model will predict the probability of the flight getting delayed.

3. First, change the directory to the `chapter10/inference` folder:

    ```
    cd chapter10/inference
    ```

4. Then, run a `curl` command to send the payload to the model. Note to change the HTTP address as per your setup:

    ```
    curl -vvvvk --header "content-type: application/json" -X
    POST -d @data.json https://flights-ontime.192.168.39.216.
    nip.io/api/v1.0/predictions; done
    ```

    Windows users may choose to use the excellent Postman application (`https://www.postman.com/`) to make an HTTP call.

5. Open the `chapter10/inference/data.json` file to see the payload that we are sending to the model. You will notice that there are two sections of the `json` payload. The first part is with the `names` key, which captures the feature columns that you have used to train the model. Notice that there is no `DELAYED` column here because the model will predict the probability of the `DELAYED` column. The second part is with the `ndarrray` key, which has the values for the feature columns. Note that the values for the categorical columns are in the original form and the inference pipeline will convert them into the categorical values before executing the model. *Figure 10.46* shows the following file:

```json
{
  "data": {
    "names": [
      "month","day","day_of_week","airline","flight_number","origin_airport","destination_airport","scheduled_time",
      "elapsed_time","air_time","distance","ORIG_CITY","ORIG_STATE","ORIG_LATITUDE","ORIG_LONGITUDE",
      "DEST_CITY","DEST_STATE","DEST_LATITUDE","DEST_LONGITUDE","scheduled_departure_hour","scheduled_departure_minute",
      "scheduled_arrival_hour","scheduled_arrival_minute"
    ],
    "ndarray": [
      [
        9,9,3,"WN",611,"LGA","DAL",89.0,82.0,64.0,447,"New York","NY",32.73356,-117.18966,
        "Dallas","TX",37.619,-122.37484,15,31,17,0
      ]
    ]
  }
}
```

Figure 10.46 – Sample payload for flights model inferencing

Now that you have successfully performed an inference call over HTTP, let's see how the information has been captured by the monitoring system.

Monitoring your model

In this last section, you will see how the platform automatically starts capturing the typical performance metrics of your model. The platform also helps you visualize the performance of the inference. The platform uses Seldon to package the model, and Seldon exposes default metrics to be captured. Seldon also allows you to write custom metrics for specific models; however, it is out of the scope of this book.

Let's start by understanding how the metrics capture and visualization work.

Understanding monitoring components

The way metrics capture works is that your model is wrapped by Seldon. Seldon then exposes the metrics to a well-defined URL endpoint, which was detailed in *Chapter 7, Model Deployment and Automation*. Prometheus harvests this information and stores it in its database. The platform's Grafana connects to Prometheus and helps you visualize the recorded metrics.

Figure 10.47 summarizes the relationship between the model and monitoring components:

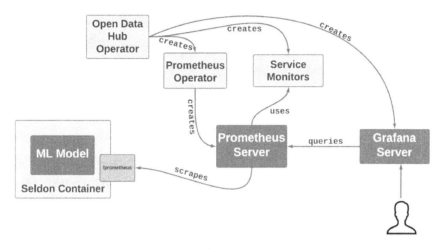

Figure 10.47 – ML platform monitoring components

Let's understand each component of this diagram:

- **Open Data Hub (ODH) Operator**: This is the base operator for our platform. Its role is to provision all the different components for your platform. We have discussed this operator in various chapters of this book and so we do not describe it in this section.

- **Prometheus Operator**: Prometheus operator is responsible for creating the Prometheus server. The ODH operator creates the Kubernetes subscriptions for the Prometheus operators. You can find the subscription file at `manifests/prometheus/base/subscription.yaml`. The following snippet shows that it uses the OLM mechanism to install the Prometheus operator:

```
manifests > prometheus > base > ! subscription.yaml > ...
1    apiVersion: operators.coreos.com/v1alpha1
2    kind: Subscription
3    metadata:
4      name: prometheus
5    spec:
6      installPlanApproval: Automatic
7      name: prometheus
8      source: community-operators-redhat
9      sourceNamespace: olm
```

Figure 10.48 – Subscription for Prometheus operator

- **Prometheus Server**: The Prometheus operator installs and configures the Prometheus server for you. The platform configures the file that directs the Prometheus operator to create the Prometheus server. You can find the file at `manifests/prometheus/base/prometheus.yaml`. The following snippet shows the file:

```
manifests > prometheus > base > ! prometheus.yaml > {} spec
     prometheus.json (prometheus.json)
1    ---
2    apiVersion: monitoring.coreos.com/v1
3    kind: Prometheus
4    metadata:
5      name: odh-monitoring
6      labels:
7        app: odh-monitoring
8        namespace: ml-workshop
9    spec:
10     replicas: 1
11     serviceAccountName: prometheus-k8s
12     securityContext: {}
13     serviceMonitorNamespaceSelector: {}
14     serviceMonitorSelector: {}
15     podMonitorSelector: {}
16     ruleSelector: {}          You, 11 hours ago • pro
```

Figure 10.49 – Prometheus server configuration

- **Service Monitors**: Service monitors are objects by which you configure the Prometheus server to find and harvest information from the running Kubernetes services and pod. The service monitors are defined by the platform, and you can find one example at `manifests/prometheus/base/prometheus.yaml`. The following snippet shows the file. Note that the configuration uses port `8000`, which is the port at which Seldon exposes the metrics information. The `selector` object defines the filter by which Prometheus will decide what pods to scrape data from:

```
manifests > prometheus > base > ! prometheus.yaml > {} spec
23    ---
24    apiVersion: monitoring.coreos.com/v1
25    kind: ServiceMonitor
26    metadata:
27      name: seldon-services
28      labels:
29        team: opendatahub
30    spec:
31      selector:
32        matchLabels:
33          app.kubernetes.io/managed-by: seldon-core
34      namespaceSelector:
35        any: true
36        # matchNames:
37        # - ml-workshop
38      endpoints:
39      - port: "8000"
40        path: /prometheus
41        interval: 10s
```

Figure 10.50 – Prometheus server monitors for Seldon pods

- **Grafana Server**: Grafana is the component that provides the visualization for the data captured by Prometheus. Grafana is preferred to create dashboards when using Prometheus and is continuously improving its Prometheus support. The platform deploys Grafana via the `manifests/grafana/base/deployment.yaml` file.

In this section, you have seen how the platform provides and wires different components to provide you with a visualization framework for your observability requirements.

Next is to configure Grafana.

Configuring Grafana and a dashboard

In this section, you will configure Grafana to connect to Prometheus and build a dashboard to visualize the model's metrics. What is a dashboard? It is a set of graphs, tables, and other visualizations of your model. You will create a dashboard for the flight model.

Note that this is a one-time configuration, and it does not need to be repeated for every model. This means that once you have a dashboard, you can use it for multiple models. Your team may create a few standard dashboards and as soon as a new model is deployed, the platform will automatically find it and make it available for monitoring.

Let's start with the configuration of the Grafana instance:

1. Log in to Grafana using `https://grafna.192.128.36.219.nip.io`. Notice that you will need to change the IP address as per your setup. On the login page, click the **Sign in With KeyCloak** button, which is at the bottom of the login window:

Figure 10.51 – Grafana login page

2. First, you will need to add a data source. A data source is a system that will provide the data that Grafana will help you visualize. The data provider in Prometheus scrapes the metrics data from your models. Select the **Configuration | Data sources** option from the left-hand menu:

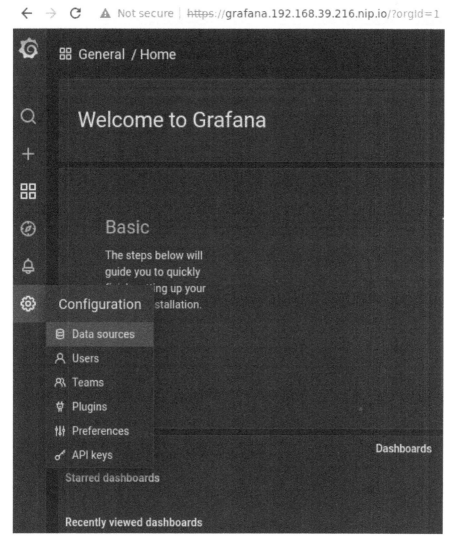

Figure 10.52 – Grafana Data sources menu option

3. Click on the **Add data source** button, as shown in the following screenshot:

Figure 10.53 – Add new Grafana data source

4. Select the data source type, which will be Prometheus for your case. You may notice that Grafana can talk to a variety of data sources, including InfluxDB and YYYY, to name a couple.

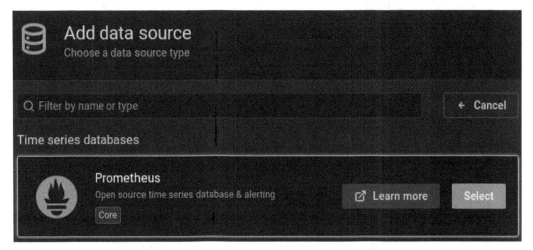

Figure 10.54 – Add new Prometheus Grafana data source

5. Now, you need to add the details for the Prometheus server. Grafana will use these details to connect and fetch data from the Prometheus server. Add the following properties in the screen mentioned:

    -   **Name:** Prometheus

    -   **URL:** http://prometheus-operated:9090

6. Then click the **Save & test** button. The URL is the location of the Prometheus service created by the platform. Because the Grafana pod will talk to the Prometheus pod using the internal Kubernetes network, this URL will be the same for your setup too:

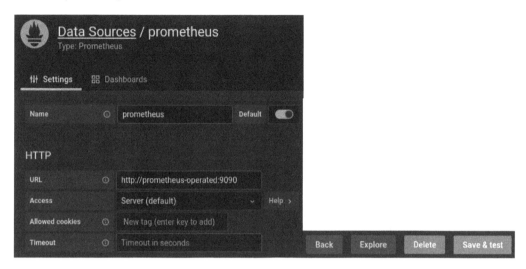

Figure 10.55 – Configuration for the Prometheus Grafana data source

You can find the `prometheus` service details by issuing the following command:

```
kubectl get service -n ml-platform | grep prometheus
```

7. After you configure Grafana to connect to Prometheus, the next step is to build the dashboard. As mentioned earlier, a dashboard is a set of visualizations, and each visualization is backed by a query. Grafana runs those queries and plots the data for you. Building dashboards is out of the scope of this book, but we have provided a dashboard that you can use. Select the **Import** option from the left-hand menu:

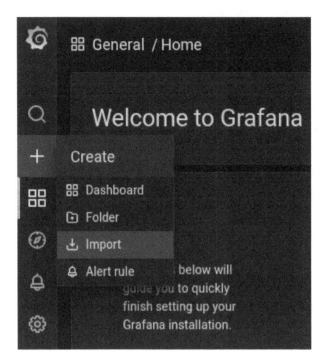

Figure 10.56 – Adding a new dashboard in Grafana

8. On the **Import** screen, copy the contents from the `chapter10/grafana-dashboard/sample-seldon-dashboard.json` file and paste it into the **Import via panel json** textbox. Click on the **Load** button to import the dashboard:

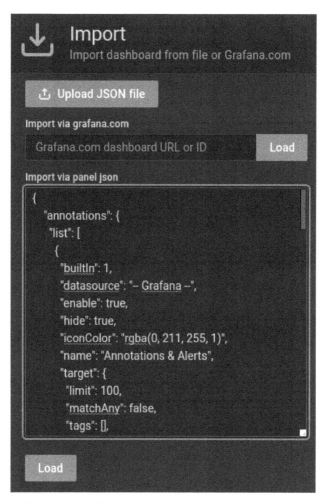

Figure 10.57 – Importing a Seldon dashboard in Grafana

9. Set the name for your imported dashboard and click on the **Import** button to complete the import process for the dashboard. You can give the name as per your liking; we have chosen the name `Flights Prediction Analytics`, as you can see in the following screenshot:

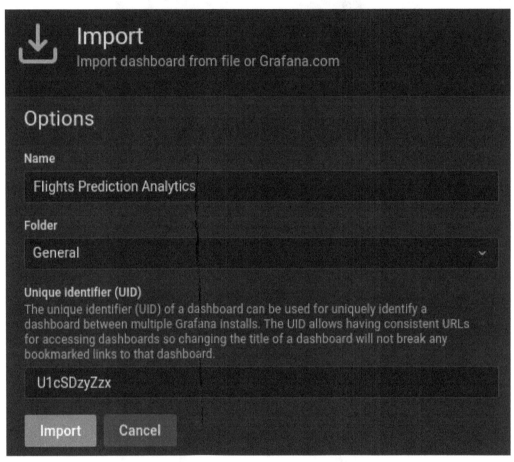

Figure 10.58 – Importing Seldon dashboard in Grafana

10. After you import the dashboard, Grafana will start displaying the dashboard immediately. You can see a few metrics such as response times, success rate, and other relative metrics for your deployed model. You may need to hit your model a few times to start populating this board. Refer to the *Calling your model* section earlier in this chapter on how to make calls to your deployed models.

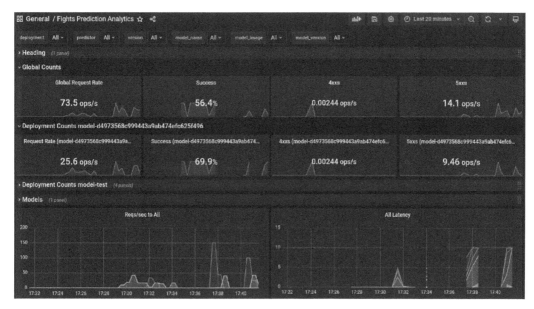

Figure 10.59 – Dashboard for Seldon models

You can see that the board captures the metrics that have been emitted by your model wrapped in Seldon. As more models get deployed, they will be available in this dashboard, and you can filter the models through the filters provided in the top bar of the dashboard.

Your flights on-time prediction service is now available for consumption. You will now work with the product development team and the website team of your organization so that they can integrate this functionality and provide a better service for your customers. Your work does not finish here; you will need to continuously see how the model is performing and bring on improvements via new data and/or optimizing your models further. The platform will help you to perform this cycle with higher velocity and continuously improve the offerings to your customers.

Summary

This was another long chapter that covered the model development and deployment life cycle for the flights on-time performance project. You have seen how the platform enables you and your team to become autonomous in EDA, model experimentation and tracking, model registry, and model deployment.

In the next chapter, we will take a step back and summarize our journey of the overall platform and how you can use it as your own solution that fits your vertical. You can use the concepts and tools to build a platform for your team and enable your business to realize the power of AI.

11
Machine Learning on Kubernetes

Throughout the chapters, you have learned about the differences between a traditional software development process and **machine learning** (**ML**). You have learned about the ML life cycle and you understand that it is pretty different from the conventional software development life cycle. We have shown you how open source software can be used to build a complete ML platform on Kubernetes. We presented to you the life cycle of ML projects, and by doing the activities, you have experienced how each phase of the project life cycle is executed.

In this chapter, we will show you some of the key ideas that we wanted to bring forth to further your knowledge on the subject. The following topics will be covered in this chapter:

- Identifying ML platform use cases
- Operationalizing ML
- Running on Kubernetes

These topics will help you decide when and where to use the ML platform that we presented in this book and help you set up the right organizational structure for running and maintaining the platform in production.

Identifying ML platform use cases

As discussed in the earlier chapters, it is imperative to understand what ML is and how it differs from other closely related disciplines, such as data analytics and data science. Data science may be required as a precursor to ML. It is instrumental in the research and exploration phase where you are unsure whether an ML algorithm can solve the problem. In the previous chapters, you have employed data science practices such as problem definitions, isolation of business metrics, and algorithm comparison. While data science is essential, there are also ML use cases that do not require as many data science activities. An example of such cases is the use of AutoML frameworks, which we will talk about in the next section.

Identifying whether ML can best solve the problem and selecting the ML platform is a bit of a chicken and egg problem. This is because, in order to be sure that an ML algorithm can best solve a certain business problem, it requires some data science work such as data exploration, and thus requires a platform to work on. If you are in this situation, your best bet is to choose an open source platform such as **Open Data Hub** (**ODH**), which we presented in this book. Because it is fully open source, there are no required commercial agreements and licenses to start installing and using the platform, and you have already seen how capable the platform is. Once you have a platform, you can then use it to initiate your research and data exploration until you can conclude whether ML is the right approach to solving the business problem or not. You can then either continue using the platform for the remainder of the project life cycle or abandon it without incurring any platform costs.

In some cases, you may already know that the business problem can be solved by ML because you have seen a similar implementation somewhere else. In such cases, choosing the ML platform we have presented is also a good option. However, you could also be in a situation where you do not have a strong data science team. You may have a few data engineers and ML engineers who understand the process of model development but are not confident about their data science skills. This is where AutoML comes into the picture as a consideration.

Considering AutoML

To define it in its simplest form, AutoML is about automatically producing ML models, with little to no data science work needed. To elaborate a bit, it is about automatic algorithm selection, automatic hyperparameter tuning, and automatic model evaluation.

AutoML technology comes as a framework or a software library that can generate an ML model from a given dataset. There are several AutoML frameworks already available on the market as of writing this book. The following list shows some of the popular AutoML frameworks currently available. There are many other AutoML frameworks not listed here, and we encourage you to explore them:

- **BigML** – An end-to-end AutoML enterprise platform sold commercially.

- **MLJAR** – An open source AutoML framework.

- **H2O.ai** – An open source full ML platform that includes an AutoML framework.

- **TPOT** – Considers itself as a data scientist assistant. It's an open source AutoML framework developed by the Computational Genetics Lab at the University of Pennsylvania.

- **MLBox** – An open source AutoML Python library.

- **Ludwig** – A toolbox featuring zero code ML model development that includes AutoML.

- **Auto-sklearn** – An open source AutoML toolkit based on scikit-learn ML libraries.

- **Auto-PyTorch** – An open source AutoML framework that features an automatic neural network architecture search. It can automatically optimize neural network architectures.

- **AutoKeras** – An open source AutoML framework based on Keras ML libraries.

It is also important to note that some of these frameworks and libraries can be used within, or in conjunction with, our ML platform or any ML platform.

Commercial platforms

Commercial vendors of ML platforms, including cloud providers, also include AutoML products and services in their portfolio. Google has Google Cloud AutoML, Microsoft has Azure Machine Learning, Amazon has Sagemaker Autopilot, and IBM has Watson Studio with AutoML and AutoAI components. However, these vendors sell their AutoML products and services as part of their ML platform product, which means you will have to use their ML platform to take advantage of the AutoML features.

ODH

You have seen how the ODH allows you to choose which components to install and it also allows you to replace one component with another by updating the `kfdef` manifest file. This adds additional flexibility as to what components you choose to be part of your platform. For example, suppose you only need JupyterHub and MLflow for your data science team to start exploring the possibility of using ML to solve your business problem. In that case, you can choose to install only these components. This will save you compute resources and, therefore, reduce cloud computing bills.

Regardless of which ML platform you choose, it is also essential that the path to operationalizing your ML platform is clearly established. This includes finding the right people who can run the platform in production and mapping the personas in the ML life cycle to the existing organization. This also includes establishing some processes and communication channels, which brings us to our next topic.

Operationalizing ML

As discussed in earlier chapters, you can enjoy the full benefits of ML in your business if your models get deployed and used in the production environment. Operationalization is more than just deploying the ML model. There are also other things that need to be addressed to have successful ML-enabled applications in production. Let's get into it.

Setting the business expectations

It is extremely important to ensure that the business stakeholders understand the risk of making business decisions using the ML model's predictions. You do not want to be in a situation where your organization fails because of ML. Zillow, a real estate company that invested a lot in ML with their product *Zestimate,* lost 500 million dollars due to incorrect price estimates of real properties. They ended up buying properties at prices set by their ML model that they eventually ended up selling for a much lower price.

ML models are not perfect; they make mistakes. The business must accept this fact and must not rely entirely on the ML model's prediction without looking at other data sources. If the business fails to accept this fact, this could lead to irreparable damages caused by wrong expectations. These damages include reputational damages, loss of trust by the business, and even regulatory fines and penalties.

Another case is that some algorithms, particularly deep learning, are not explainable. It must be communicated to the business because, in some cases, an explainable algorithm may be required for regulatory purposes. Some regulators may need you to explain the reason behind the business decision. For example, suppose an ML model decided that a new bank customer is not a risky individual and it turned out to be a black-listed or sanctioned individual by some regulators; the financial organization may need to explain the reasoning behind this decision to the regulators during the investigation and the post-mortem analysis. Or, even worse, the organization could get fined millions of dollars.

Avoid over-promising results to the business. IBM Watson had the idea that ML could diagnose cancer by making sense of diagnostic data from several medical institutions and potentially assisting, or even replacing, doctors in performing a more reliable cancer diagnosis in the future. This has gained a lot of attention, and many organizations invested in the idea. However, it turned out to be a very difficult task. It did not only result in losses, but it also somehow damaged the brand.

To summarize, before deciding whether to use ML models to predict business decisions, make sure that the business understands the risks and consequences if the model does not behave as expected. Set the expectations right. Be transparent about what is possible and what is hard. Some ML models may be able to replace a human in a particular business process, but not all ML models will achieve superhuman abilities.

Dealing with dirty real-world data

The data you used for model training comes as prepared datasets tested in a controlled environment. However, this is not the case in the real-world setting. After your model gets deployed to production, you must expect dirty data. You may receive wrongly structured data, and most of the data is new and has never been seen by the model during training. To ensure that your model is fit for production, avoid overfitting, and test the model thoroughly with datasets that are as close as the ones it will see in production. If possible, use data augmentation techniques or even manufactured data to simulate production scenarios. For example, a model that works well in diagnosing a patient utilizing chest X-ray scans may work well in one clinic, but it may not work in another clinic using older medical equipment. There is a real story behind this, and the reason it did not work was that the X-ray scanners generated scans that showed dust particles present in the machine's sensors.

To summarize, avoid overfitting. Have a solid data cleaning process as part of your inference pipeline. Prepare for the worst possible input data by having suitable datasets from various sources. Be ready when your model does not return what is expected of it.

Dealing with incorrect results

Imagine you have a credit card fraud detection and it marks a routine transaction as fraudulent. There could be many reasons for this, such as your model may not be aware of higher-than-normal spending during Christmas. You need the capability to investigate such scenarios and that's why it is crucial to have logging in place. This will allow you to recall the model's answer to a particular question thrown to it in production. You will need this to investigate model issues.

When this happens, you must be prepared to face the consequences of the wrong information your model returned. But also, you must be able to address the erroneous result in the future by updating the model with new sets of data from time to time. You must also have the ability to track the model's performance over time. You have seen in the previous chapter how monitoring is done. The change in model performance over time is also called a **drift**. There are two kinds of drift. **Data drift** happens when the model starts receiving new types of data that it has not been trained on. For example, an insurance fraud detection model worked well until it started seeing new data that included a new insurance product that the model hadn't seen before. In this case, the model will not produce a reliable result. In other words, your model performance has degraded. Another example is that your model was trained on a certain demographic or age group, and then suddenly a new age group started appearing. Similarly, there is a higher chance that the ML model will return an unreliable result. **Concept drift** is when the functional relationship between the input data and the label has changed. For example, in a fraud detection model, a transaction that was not previously considered fraudulent is now labeled as fraudulent or anomalous according to the new regulations. This means the model will produce more false-negative results, which renders the model unreliable.

In these scenarios, you must have a process set for addressing these problems. You must have a process for when to manually retrain the model, or even automatically retrain the model when it detects a drift. You may also want to implement anomaly detection in the input data. This ensures that your model only gives up results if the input data make sense. This avoids abuse or attacks on the model as well. These automation requirements can be integrated as part of your continuous integration and deployment pipelines.

Maintaining continuous delivery

You have seen how to run model builds and model deployments in the platform manually. You have also seen how to automate the deployment workflow using Airflow. Although the data scientists or ML engineers in the team can manually perform or trigger such operations, in the real world, you will also need someone or a team to maintain these pipelines to make sure they are always working. You may want to have a dedicated platform team to maintain the underlying platform that executes the pipelines, or you may assign this responsibility to the data engineering team. Whatever approach you choose, the important thing is that someone must be responsible for ensuring that the deployment pipelines are always working.

Although the ODH operator completely manages the ML platform, you will still need someone responsible for maintaining it. Ensure that the Kubernetes operators are up to date. Apply security patches whenever necessary.

For some critical workloads, you may not be able to deploy to production automatically. There will be manual approvals required before you can ship updates to a model in production. In this case, you need to establish this approval workflow by either embedding this process into the platform or through mutual agreement with manual approval processes. Nevertheless, the objective is to have someone accountable for maintaining continuous delivery services.

In summary, continuous delivery must always work so that the model development life cycle can have a faster feedback cycle. Also, if drift is detected, you will always have a ready-to-go delivery pipeline that can ship a more up-to-date version of the model.

Managing security

Security is another critical area to focus on when operationalizing ML projects. You have seen in the preceding chapters that the ML platform can be secured by using **OpenID Connect** (**OIDC**) or **OAuth2**, a standard authentication mechanism. Different platform components can utilize the same authentication mechanism for a more seamless user experience. You have used an open source tool called Keycloak, an industry-standard implementation of the **identity and access management** (**IAM**) system that mainly supports OIDC, **Security Assertion Markup Language** (**SAML**), and more. The Seldon Core API allows the REST-exposed ML models to be protected behind the same authentication mechanism. Refer to the Seldon Core documentation for more details.

To summarize, the ML platform must be protected by an authentication mechanism, preferably OIDC. This also allows for the implementation of **single sign-on** (**SSO**). Additionally, you also need to protect your deployed models to ensure that only the intended audiences can access your ML models. And finally, there must be someone responsible for maintaining the Keycloak instance that your platform uses and someone, or a team, managing the access to the platform resources.

Adhering to compliance policies

In some business settings, compliance is at the center of the operation. Financial institutions have a whole department managing compliance. These compliance rules typically come from the regulatory bodies that oversee the financial institution's operations. Depending on which country your ML platform will be used and hosted in, regulatory policies may prevent you from moving data out of the on-premises data centers. Or, there could be a requirement for encrypting data at rest.

The good news is that your platform is flexible enough to be configured for such compliance measures. It can run on-premises or in any cloud provider, thanks to Kubernetes. You can also run the ML platform in the cloud while having the storage on-premises or take advantage of hybrid-cloud strategies.

Another thing is that each of the components in the platform is replaceable and pluggable. For example, instead of using a dedicated instance of Keycloak, you could use an existing regulator-approved OIDC provider.

Compliance could often become an impediment in progressing with ML projects. If you plan to use a commercial platform rather than the one you built in this book, always consider the compliance or regulatory requirements before deciding. Some commercial platforms in the cloud may not be able to comply with data sovereignty, especially in countries where the major cloud providers do not yet have a local data center.

In other words, always consider compliance requirements when planning for the architecture of your ML platform.

Applying governance

After taking into account the preceding considerations, another important area that needs to be cleared out to operationalize your ML platform is **governance**. This is where you will design the organizational structure, roles and responsibilities, collaboration model, and escalation points. The authors advocate for a more cross-functional team with very high collaboration levels. However, this is not always possible in the real world. There are organizations with very well-defined hierarchies and silos that refuse to change the way things are. If you are in this type of organization, you may face several hurdles in implementing the ML platform we have presented here.

One of the platform's main features is that it is a self-service platform. It allows data scientists, ML engineers, and data engineers to spin up their notebook servers and Spark clusters. However, this will also lead to less predictable cloud billings or operating costs. If you are the data architect of the project, part of your job is to convince the leadership team and the platform teams to trust their data scientists and ML engineers.

Ideally, the best way to design the organizational structure around the ML project is to have a platform team. This team is responsible for running the ML platform. This team then acts as a service provider to the data and application teams, also called the **stream-aligned teams**, in a **software as a service (SaaS)** model. The platform team's objective is to ensure that the stream-aligned teams can perform their work on the platform as smoothly and as quickly as possible. The data science and data engineering teams can be the stream-aligned teams, and they are the main users of the platform and the main customers of the platform team. The DevSecOps or DevOps teams may sit together in the same organizational unit, as the platform team provides DevOps services to the stream-aligned teams. *Figure 11.1* shows an example of an organizational structure that you could implement to run an ML project using the Team Topologies notation:

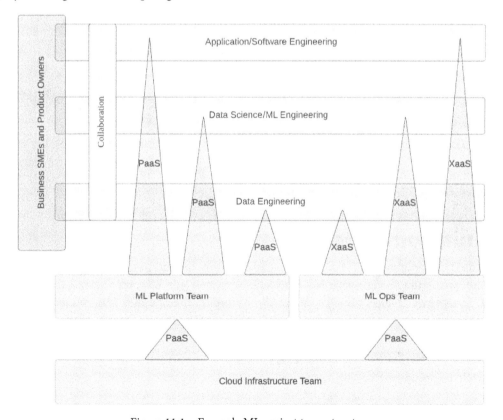

Figure 11.1 – Example ML project team structure

In *Figure 11.1*, there are a total of three stream-aligned teams, namely, the data science team, the data engineering team, and the software engineering team. All three stream-aligned teams are collaborating with each other with the objective of delivering an ML-enabled application in production. There are also three platform teams. The cloud infrastructure team is providing a cloud **platform as a service** (**PaaS**) to the two other platform teams: the ML platform team and the MLOps team. Both the ML platform team and the MLOps team are providing ML PaaS and MLOps as a service to all the three stream-aligned teams. The purple box represents an enabling team. This is where the SMEs and product owners sit. This team enables and provides support to all the stream-aligned teams.

You must take note that this is just an example; you may want to combine the ML platform team and MLOps team together, or the data science and data engineering teams, and that's perfectly okay.

If you want to learn more about this type of organizational design notation, you may want to read about Team Topologies.

We can summarize as follows:

- Use the ML life cycle diagram that you have seen in *Figure 2.7* in *Chapter 2, Understanding MLOps*, to map the current organizational structure of your teams.

- Communicate the roles and responsibilities clearly.

- Set the collaboration channels and feedback points, such as design spike meetings and chatgroups.

Suppose you cannot break the silos; set up regular meetings between the silos and establish a more streamlined handover process. However, if you want to take advantage of the full potential of the ML platform, we strongly recommend that you form a cross-functional and self-organizing team to deliver your ML project.

Running on Kubernetes

Using the ODH operator, the ML platform truly unlocks the full potential of Kubernetes as the infrastructure layer of your ML platform. The **Operator Lifecycle Management** (**OLM**) framework enables the ODH operator to simplify the operation and maintenance of the ML platform. Almost all operational work is done in a Kubernetes-native way, and you can even spin up multiple ML platforms with a few clicks. Kubernetes and the OLM also allow you to implement the **Platform as Code** (**PaC**) approach, enabling you to implement GitOps practices.

The ML platform you've seen in this book works well with vanilla Kubernetes instances or any other flavors of Kubernetes or even a Kubernetes-based platform. In fact, the original ODH repository was mainly designed and built for Red Hat OpenShift.

Avoiding vendor lock-ins

Kubernetes protects you from vendor lock-ins. Because of the extra layer of containerization and container orchestration, all your workloads do not run directly on the infrastructure layer but through containers. This allows the ML platform to be hosted in any capable infrastructure. Whether on-premises or in the cloud, the operations will be the same. This also allows you to seamlessly switch to a different cloud provider when needed. This is one of the advantages of using this ML platform when compared to the commercial platforms provided by cloud vendors. You are not subject to vendor lock-in.

For example, if you use Azure ML as your platform of choice, you will be stuck with using Azure as your infrastructure provider. You will not be able to move your entire ML project to another cloud vendor without changing the platform and deployment architecture. In other words, the cost of switching to a different cloud vendor is so high that you are basically stuck with the original vendor.

Considering other Kubernetes platforms

It is not mandatory for this ML platform to run on the vanilla Kubernetes platform only. As mentioned in the previous section, the original ODH was designed to run on Red Hat OpenShift, whereas in this book, you managed to make it run on minikube, a single-node vanilla Kubernetes.

There are many other Kubernetes platforms out there, including those provided by the major cloud providers. The following list includes the most common ones in no particular order, but other emerging Kubernetes-based platforms have just entered the market or are either in beta or in development as of this writing:

- **Kubernetes**
- **Red Hat OpenShift Container Platform (OCP)**
- **Google Kubernetes Engine (GKE)**
- **Amazon Elastic Kubernetes Engine (EKS)**
- **Azure Kubernetes Service (AKS)**
- **VMware Tanzu**
- **Docker Enterprise Edition (Docker EE)**

Although we have tested this platform in Kubernetes and Red Hat OpenShift, the ML platform that you built in minikube can also be built in any of the above Kubernetes platforms, and others. But, what about in the future? Where is ODH heading?

Roadmap

ODH is an active open source project primarily maintained by Red Hat, the largest open source company in the world. ODH will keep getting updated to bring more and more features to the product. However, because the ML and MLOps space is also relatively new and still evolving, it is not unnatural to see significant changes and pivots in the project over time.

As of writing this book, the next version of ODH includes the following changes (as shown in *Figure 11.2*):

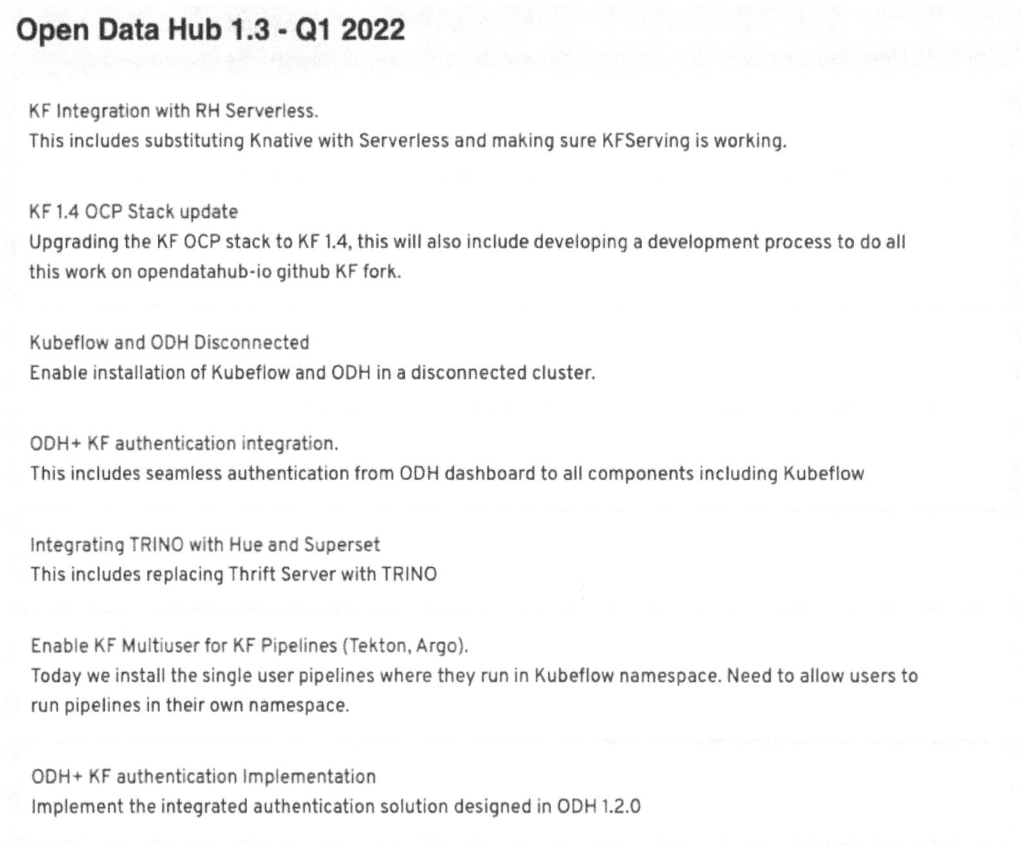

Open Data Hub 1.3 - Q1 2022

KF Integration with RH Serverless.
This includes substituting Knative with Serverless and making sure KFServing is working.

KF 1.4 OCP Stack update
Upgrading the KF OCP stack to KF 1.4, this will also include developing a development process to do all this work on opendatahub-io github KF fork.

Kubeflow and ODH Disconnected
Enable installation of Kubeflow and ODH in a disconnected cluster.

ODH+ KF authentication integration.
This includes seamless authentication from ODH dashboard to all components including Kubeflow

Integrating TRINO with Hue and Superset
This includes replacing Thrift Server with TRINO

Enable KF Multiuser for KF Pipelines (Tekton, Argo).
Today we install the single user pipelines where they run in Kubeflow namespace. Need to allow users to run pipelines in their own namespace.

ODH+ KF authentication Implementation
Implement the integrated authentication solution designed in ODH 1.2.0

Figure 11.2 – ODH's next release

There are other features of ODH that you have not yet explored because they are more geared toward data engineering and the data analytics space. One example is data virtualization and visualization using Trino and Superset. If you want to learn more about these features, you can explore them in the same ML platform you built by simply updating the `kfdef` file to include Trino and Superset as components of your ML platform. You will find some examples of these `kfdef` files in the ODH GitHub project.

You can look for future roadmaps of ODH at the following URL: `https://opendatahub.io/docs/roadmap/future.html`.

In the future, there could be another open source ML platform project that will surface on the market. Keep an open mind, and never stop exploring other open source projects.

Summary

The knowledge that you have gained in this book about ML, data science and data engineering, MLOps, and the ML life cycle applies to any other ML platforms as well. You have not only gained important insights and knowledge about running ML projects in Kubernetes but also gained the experience of building the platform from scratch. In the later chapters, you were able to gain hands-on experience and wear the hats of a data engineer, data scientist, and MLOps engineer.

While writing this book, we realized that the subject is vast and that going deep into each of the topics covered in the book may be too much for some. Although we have touched upon most of the components of the ML platform, there is still a lot more to learn about each of the components, especially Seldon Core, Apache Spark, and Apache Airflow. To further your knowledge of these applications, we recommend going through the official documentation pages.

ML, AI, and MLOps are still evolving. On the other hand, even though Kubernetes is almost 8 years old, it is still relatively new to most enterprise organizations. Because of this, most professionals in this space are still learning, while at the same time establishing new standards.

Keep yourself updated on the latest ML and Kubernetes trends. You already have enough knowledge to advance your learning in this subject on your own.

Further reading

- *Seldon core documentation*: `https://docs.seldon.io/projects/seldon-core/en/latest/index.html`

- *Team topologies*: `https://teamtopologies.com`

- *Open Data Hub*: `https://opendatahub.io`

Index

Packt.com

Subscribe to our online digital library for full access to over 7,000 books and videos, as well as industry leading tools to help you plan your personal development and advance your career. For more information, please visit our website.

Why subscribe?

- Spend less time learning and more time coding with practical eBooks and Videos from over 4,000 industry professionals

- Improve your learning with Skill Plans built especially for you

- Get a free eBook or video every month

- Fully searchable for easy access to vital information

- Copy and paste, print, and bookmark content

Did you know that Packt offers eBook versions of every book published, with PDF and ePub files available? You can upgrade to the eBook version at packt.com and as a print book customer, you are entitled to a discount on the eBook copy. Get in touch with us at customercare@packtpub.com for more details.

At www.packt.com, you can also read a collection of free technical articles, sign up for a range of free newsletters, and receive exclusive discounts and offers on Packt books and eBooks.

Other Books You May Enjoy

If you enjoyed this book, you may be interested in these other books by Packt:

The Kubernetes Workshop

Zachary Arnold, Sahil Dua, Wei Huang, Faisal Masood, Melony Qin, Mohammed Abu Taleb

ISBN: 978-1-83882-075-6

- Get to grips with the fundamentals of Kubernetes and its terminology
- Share or store data in different containers running in the same pod
- Create a container image from an image definition manifest
- Construct a Kubernetes-aware continuous integration (CI) pipeline for deployments
- Attract traffic to your app using Kubernetes ingress
- Build and deploy your own admission controller

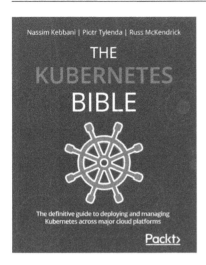

The Kubernetes Bible

Nassim Kebbani, Piotr Tylenda, Russ McKendrick

ISBN: 978-1-83882-769-4

- Manage containerized applications with Kubernetes
- Understand Kubernetes architecture and the responsibilities of each component
- Set up Kubernetes on Amazon Elastic Kubernetes Service, Google Kubernetes Engine, and Microsoft Azure Kubernetes Service
- Deploy cloud applications such as Prometheus and Elasticsearch using Helm charts
- Discover advanced techniques for Pod scheduling and auto-scaling the cluster
- Understand possible approaches to traffic routing in Kubernetes

Packt is searching for authors like you

If you're interested in becoming an author for Packt, please visit `authors.packtpub.com` and apply today. We have worked with thousands of developers and tech professionals, just like you, to help them share their insight with the global tech community. You can make a general application, apply for a specific hot topic that we are recruiting an author for, or submit your own idea.

Share Your Thoughts

Now you've finished *Machine Learning on Kubernetes*, we'd love to hear your thoughts! Scan the QR code below to go straight to the Amazon review page for this book and share your feedback or leave a review on the site that you purchased it from.

https://packt.link/r/1-803-24180-2

Your review is important to us and the tech community and will help us make sure we're delivering excellent quality content.

www.ingramcontent.com/pod-product-compliance
Lightning Source LLC
Chambersburg PA
CBHW060923060326
40690CB00041B/3000